THROUGH THE TELESCOPE

A GUIDE FOR THE AMATEUR ASTRONOMER

Revised and Updated

Patricia L. Barnes-Svarney
Michael R. Porcellino

McGraw-Hill

New York   San Francisco   Washington, D.C.   Auckland   Bogotá
Caracas   Lisbon   London   Madrid   Mexico City   Milan
Montreal   New Delhi   San Juan   Singapore
Sydney   Tokyo   Toronto

McGraw-Hill

A Division of The McGraw·Hill Companies

3 4 5 6 7 8 9 0   DOC/DOC   0 9 8 7 6 5 4 3 2 1

ISBN 0-07-134804-2

This book was designed, edited, and set in Granjon and Charme by TopDesk Publishers' Group. *Printed and Bound by* R.R. Donnelley & Sons Company.

McGraw-Hill books are available at special quantity discounts to use as premiums and sales promotions, or for use in corporate training programs. For more information, please write to the Director of Special Sales, McGraw-Hill, Professional Publishing, Two Penn Plaza, New York, NY 10121-2298. Or contact your local bookstore.

 This book is printed on recycled, acid-free paper containing a minimum of 50% recycled, de-inked fiber.

CONTENTS

INTRODUCTION

Welcome to the wonderful universe of amateur astronomy. Amateur astronomy is something that anyone can enjoy. All that is needed are clear skies and a dark place to start you on the road to the stars. While no one can supply clear skies, this book can start you on your way.

I wrote this book with two kinds of amateurs in mind. First, it is written for the beginner who is thinking of getting, or has just gotten, a telescope and wants to learn just what it can show them. If you find yourself in this group, you have to take some time and prepare yourself. A big mistake that many people make is to rush out and get a telescope before they are ready to use it.

When is a person ready for a telescope? You must have a working knowledge of the constellations and the star figures that you are going to wander through with your telescope. You wouldn't leave home for a long trip without some idea of where you were going, would you? Well, that's just what you are doing without a knowledge of the constellations. If you don't know where Epsilon Bootes is, then you are never going to be able to find the variable star W Bootes. So take some time and familiarize yourself with the stars. When you know your way around the stars that you can see with your naked eye, then you are ready for a telescope and the stars that you cannot see!

The other group I wrote this book for includes those with a small telescope. They have reached a point where they can easily find their way around the skies, and they have invested in a telescope but they don't know where to point it. This book will serve as a guide to using that instrument. Many of today's observing guides tend to pass over these amateurs, an omission I find unfortunate, and these small scope users feel they have been cheated.

Unfortunately, in this age of 17.5-inch dobsonians, too many of today's amateurs look down on the under 4-inch telescope like a poor relative. But if you ask 100 amateur astronomers how big their first telescope was, I'll wager that 85 percent of them will say it was less than 4 inches. These amateurs and the writers of observation guides have forgotten their "roots." Since that first telescope they have moved up to the huge "light buckets" that have become all the rage, and they now have little use for the small telescope. Everyone in amateur astronomy should remember that they started with some kind of telescope. In many cases it was a small one.

To meet the needs of both these segments of the astronomical community, I wrote this book as a survey of the wide variety of experiences that an amateur with a telescope can enjoy. I've tried to give an overview of the equipment needed by the observer and some thoughts on its selection. I've also tried to show that an amateur with a small telescope can make a contribution to science. Sure, some activities require a large-aperture instrument, like supernova searches, but other areas just as important to science require only a small instrument.

Whether you are just entering the wonderful universe of amateur astronomy or have been in it for awhile, I want you to remember one thing above all else. Amateur astronomy should be fun! This is not a contest to see who can spot the most features on the planet Mars or estimate 100 variables in one night. It is an immensely enjoyable, relaxing way to spend your time, and if you can make some contribution to science along the way that's great, but it is not a requirement.

So enjoy the stars, enjoy your telescope, and most of all enjoy this book.

Clear skies to you all!

—Michael Porcellino, 1989

In the ten years since this book was first written, astronomy has grown by leaps and bounds. There was Shoemaker-Levy-9, the first body (a comet) we ever viewed striking another body (the gas giant Jupiter). The astounding images sent back by the Hubble Space Telescope. The bright comet Hale-Bopp in 1997. The "fleet" of Martian probes, not to mention craft sent to the Moon, Jupiter, Venus, Saturn, and to a near-Earth asteroid.

Amateur astronomy has grown right along with these wonders. The Internet has produced a way for national and worldwide amateur groups to exchange information. The huge network—virtually a giant library—has sparked the interest of more than one future astronomer. And the ability to rapidly report on objects in the nighttime skies also has improved our disposition.

Technology also has enhanced our lives when it comes to watching the heavens. Lenses have improved, along with various new eyepieces with better fields of view. There are CCD telescopes that use technology once the purview of military satellite technology or a few large telescopes. Even small satellite dishes are being modified to allow the amateur to seek out the radio-noise of the skies.

But underneath it all, there are still those of us who want to wander the stars at a slower pace—to know what's out there right now. In other words, to watch, not to get caught up too much in the hype. That's what this book for the amateur astronomer—especially for the beginner and intermediate watches—offers.

In my work and travels as an astronomer (not to mention my many interviews with amateur astronomers), I've met some grand people: long nights with comet hunters as they search the skies for elusive fuzzy spots; sun watchers, counting sunspots before and after solar cycle peaks; an amateur astronomer seeking extrasolar planets in his backyard with a method he developed—and not with the advantage of a huge telescope; the amateur who had an experiment on the Hubble Space Telescope. And even a young woman in her teens who worked out the shape of the first asteroid ever visited (Gaspra) even before the spacecraft arrived.

The list goes on. These amateurs—including myself—have a love of the universe that goes much farther than the naked eye or their telescopes can see. It's a passion—a feeling that you want to discover and uncover something wondrous in the universe. We want to satiate our own curiosity—and to perhaps contribute in some way along our astronomical journey.

It is to these spirited people that this book is written—and all the future amateur astronomers who follow in their footsteps. The universe is yours for the taking.

—Patricia Barnes-Svarney, 1999

ACKNOWLEDGMENTS

A book is never the work of only one person. By meeting and talking to a great number of people, I have been able to garner bits of information that, in one form or another, eventually made their way into this book. Over the course of writing this book, I have had the pleasure and the honor of meeting, talking to, and corresponding with amateurs from around the country. In their way, however small, many people have contributed to the completion of this book.

I would like to thank Jim Seevers and the staff of the Adler Planetarium of Chicago for allowing me to use their excellent library during the research of this book. I would especially like to thank librarians Pamela Hovan and Evelyn Natividad for putting up with me while I rummaged through books at the back table.

I would like to thank Jeffrey Hunt of Cosmic Connections for reading most of this manuscript and making invaluable comments about its content.

The members of the Chicago Astronomical Society, whether they know it or not, helped in many aspects of the work with their comments. I would especially like to thank Jim Carroll, John Jones, Robert Stalzer, and Kathy Piorkowski for their contributions and suggestions.

Other members of the amateur astronomical community from around the country also helped with this project. I want to thank Chet Beck, Paul Bock, Robert Browning, Tom Johnson, and Jim Senk for their letters and comments, which helped develop the book from concept to completion.

Thanks also to Pat Shand and the staff of Lick Observatory for some of the photographs found in this book. Jeff Beish, Don Parker, Jim Phillips, and the other members of the Association of Lunar and Planetary Observers (ALPO) helped, as did Janet Mattei and the American Association of Variable Star Observers (AAVSO). I thank you all.

I want to extend a special thanks to Dave and Marion Bachtell, Barbara Lux, Dan Joyce, Lee Keith, Walter Piorkowski, Dan Troiani, Suresh Sreenivasan, Mark Stauffer, and Linda Warren for the drawings, photographs, and ideas they contributed to this book. Without their help it would never have been completed.

Any errors or omissions found in this work are mine and mine alone.

Finally, I want to thank Roland Phelps, my editor at Tab Books, for putting up with me during the course of this project.

—Michael Porcellino, 1989

As Mike mentioned, no one person writes a book, and I'd like to add my own acknowledgements: I'd also like to thank Thomas Svarney for his help with research; Don Parker and Don Troiani of ALPO; Janet Mattei of AAVSO; Dr. Alan Hale, co-discoverer of comet Hale-Bopp and head of the Southwest Institute for Space Research; Reverend Robert Evans of Australia; Dr. Peter Thomas, Space Science, Cornell University; the Lick Observatory; the Kopernik Observatory in Vestal, New York; and the people I've met along the way—both amateur and professional astronomers who will never give up exploring the universe.

In particular, I'd like to thank my editor, Griffin Hansbury, for his fine editing and good humor. And also, of course, I'd like to thank my parents, William and Helen Barnes, who for years have nurtured and encouraged me in my writing and astronomical adventures. I couldn't have done it without them.

—Patricia Barnes-Svarney, 1999
e-mail: *svarney@ibm.net*
home page: *http://www.sff.net/people/P.Barnes-Svarney*

1

THE AMATEUR
IN ASTRONOMY

Astronomy offers a pleasure that follows the law of increasing rather than diminishing returns. The more you develop a thirst for it, the greater your return.

WILLIAM T. OLCOTT

There is something almost mystical about sitting under a crystal clear night sky while the strains of Bach or the Beatles or Spyro Gyra play on your portable stereo- or as nighttime silence surrounds you. Sitting and studying the shapes of the constellations as they wheel across the black velvet dome of the sky is both exciting and relaxing. This is the realm of the amateur astronomer.

There are many definitions of the word *amateur*. A few listed by Webster's Dictionary include "dilettante," "nonprofessional," and a personal favorite, "one who is unskilled." These definitions might be adequate for an entry in Webster's, but they hardly describe the "amateur" in amateur astronomer. These definitions forget that *amateur* comes from the Latin word *amator*, meaning "one who loves." That describes the amateur astronomer; it describes one who loves astronomy (Fig. 1-1).

THE AMATEUR TRADITION IN ASTRONOMY

During most of the eighteenth century, astronomy was a practical science. The job of the astronomer was simple: to produce information that would be useful in the exploration of the world. So the astronomer tried to determine the positions of stars as accurately as possible. This kind of information was just the thing a sailor or surveyor needed to know to figure out his position on the Earth.

Until the late eighteenth century, positional astronomy was the order of the day at the observatory. In many observatories it was forbidden for the staff to use the

FIGURE 1-1   *The Earth and Moon from the Galileo spacecraft—something only viewed from space, but it shows our place in the solar system. (NASA)*

telescopes for anything but the measurement of star positions. It was also a time when many who held the title of astronomer worked their way up the ladder through the boring jobs of calculating ephemerides and copying charts.

This situation began to change in the mid-1700s with the invention of the achromatic lens system for refracting telescopes by John Dolland. Apprentice astronomers, taken with the images of the stars during their stints at the transit telescope, began to get curious. What would the Moon look like with this instrument? They would risk the wrath of their employers and sneak a peek. Soon, even their employers were peeking into the universe. Others began experimenting by casting metal mirrors and constructing instruments along the lines suggested by scientist Sir Isaac Newton. Overall, these telescopes were required to do nothing more than show their owners the universe.

The man credited with the invention of observational astronomy, Sir William Herschel, was by trade a musician (although the first person to use a telescope for astronomical observation was Galileo Galilei in 1609). Each night, Herschel would retire with a glass of milk and two books: one on harmonics, the study of musical sounds, and the other on astronomy. As his interest in astronomy was piqued, he began constructing simple telescopes from spectacle lenses. In 1774, he began constructing large reflecting telescopes and using them for a systematic study of the sky—for no other purpose than to see what was out there. Observational astronomy was born.

In the United States, no one exemplified the amateur tradition in astronomy better than Sherburne Wesley Burnham. Born in Vermont in 1838, Burnham was trained as a shorthand writer. He served in the Civil War as a court reporter with the Union forces occupying New Orleans, and later as the Chief Clerk of the Circuit Court in Chicago. He would work 8 hours at his job and then stay up far into the night studying double stars from the backyard of his home in Chicago. Using a 6-inch refractor built for him by the Massachusetts firm of Alvan Clark & Sons, Burnham discovered and studied hundreds of double stars. His publication of his observations lead to recognition by scientific societies around the world and entry into some of the world's most prestigious observatories. With the exception of a two-year stint on the staff of Lick Observatory, all of Burnham's work, until his death in 1921, was done as an amateur.

In the twentieth century, the amateur tradition has been represented by a number of observers. Probably best known among them was Leslie Peltier of Delphos, Ohio. Peltier began his love of the heavens as a boy and carried the torch until his death in 1980. Called the "world's best nonprofessional astronomer" by Harvard College Observatory director Harlow Shapley, Pettier's career spanned almost 70 years. During this time he discovered 12 comets, contributed over 100,000 brightness estimates of variable stars, and made numerous other accomplishments in the field of astronomy.

Today, the tradition continues. A doctor in South Carolina spends his time away from the hospital studying features on the Moon known as domes in an attempt to understand lunar volcanic action. A postal worker in Chicago rises well before dawn to study Mars and contribute to our growing knowledge of that planet's weather. A housewife in Virginia watches the Sun with her special telescope, keeping track of sunspot activity every sunny day. A chemist in Baton Rouge meticulously measures the distance between double stars in an attempt to refine the knowledge gained about them over the past 200 years.

WHY AMATEUR ASTRONOMY?

People become interested in astronomy for a variety of reasons. Ask a hundred amateur astronomers how they became interested and you will probably get a hundred different answers. Many will admit that using a small telescope they received as a child began a lifelong love affair with the stars. For others it was a trip to a planetarium or an event sponsored by a local astronomy club that sparked their interest.

Some were simply fascinated by the patterns the stars created in the night sky. After spending some time in the local library reading up on stars, constellations, and astronomy, they soon found themselves able to identify such star groups as Orion, Lyra, and Leo. Still others owe their interest in astronomy to their children. Helping with a science project has given many a parent a taste for science in general and astronomy in particular.

Regardless of how your interest in astronomy was sparked, there are a number of ways you can get more involved with astronomy. First, you can simply experience the sky, which has fascinated people for hundreds of centuries. If a pair of binoculars is around the house, you can use them to turn toward the stars. Once you have a taste of what lies beyond the grasp of the naked eye, you will realize your appetite has only been whetted. Finally, you might buy or make a telescope. Each of these stages is a learning experience and if progressed through properly, you may eventually make significant contributions to science.

You could soon find yourself rising at 3:00 A.M. to get a good view of Mars or Jupiter. You might find yourself thrilled by the changing brightness of a distant star and record those changes at every opportunity. The rolling line of sunrise on the Moon may catch your fancy, and you might find yourself watching it night after night—until you realize the distant surface of the Moon is as familiar to you as your own backyard.

CONTRIBUTIONS FROM THE AMATEUR

Amateur astronomers fill a void in today's science. Unfortunately, budget restrictions at many of the major observatories make some types of astronomical work uneconomical. The amateur is relied upon to fill the void. A prime example is the study of the planets. With such missions to Mars as the Viking, Pathfinder, and Mars Global Surveyor, the Voyager missions to the outer planets, Galileo to Jupiter, and Cassini to Saturn, many astronomers felt the days of Earth-based planetary observation were over.

However, despite their mind-boggling results, these missions last only a finite time. If we are going to launch and carry out a manned mission to Mars, for example, we have to have a better understanding of Martian weather. It doesn't do any good to mount an expensive manned mission that will last two or more years and travel over 120 million miles only to have the spacecraft orbit the red planet, unable to land due to a massive dust storm. Much of our knowledge of Martian weather, including the growth and predictions of major dust storms, has come from the work of amateur astronomers such as executive director Donald Parker, Daniel Troiani, and other members of the Association of Lunar and Planetary Observers (ALPO).

Another example is the internal constitution of the stars, long the purview of professional astronomers: Such studies have also felt the impact of the amateur astronomer. Now that observational time on large telescopes is severely limited, amateurs are contributing observations of the fluctuations in brightness of stars. This information helps scientists determine what makes a star shine. Since its foundation in 1911, the members of the American Association of Variable Star Observers (AAVSO) have contributed over 5 million observations of stellar brightness. The role played by the members of AAVSO has also grown in importance with the introduction of sophisticated satellites studying stars from Earth's orbit. Often amateur observations provided by AAVSO enable scientists to correlate an event observed by their satellites with actual visual observations.

There is nothing that says you must, as an amateur astronomer, make such observations. The beauty of amateur astronomy is that you can make contributions like these if you want to—not because you must.

ONE AMATEUR'S EXPERIENCE

Just what kind of contribution can a single amateur make to the world of astronomy? At first glance it would seem hopeless to try to compete with the huge telescopes and complex instruments of the professional, but have faith, it is possible.

Dan Troiani first became interested in astronomy in 1971 when he visited Chicago's Adler Planetarium with his future wife, Kathy. He was fascinated with the sky show and the wide variety of astronomical exhibits at Adler. He bean to read about astronomy and became more deeply interested. He also became an avid *Star Trek* fan, but found he was more interested in the "science" than the fiction. "I realized after a short time," he said, "that the science of astronomy was stranger and weirder than fiction could ever be."

In his wide readings, Dan was taken by the amount of detail visible on the planetary drawings he saw. "Could I do that?" he asked himself. The skies over a large city like Chicago are not conducive to many types of astronomical work. It became apparent to Dan during the course of his research that the Moon and planets were the perfect objects for a city astronomer to study.

In the 1970s, before the revolution in telescope design brought about by lightweight mirrors and the dobsonian mount, a 10- or 12.5-inch newtonian reflector was considered a large telescope. Dan decided that a 10-inch reflector would be the perfect tool for a serious program of planetary study. So in March 1977, he ordered a 10-inch mirror from Cave Optical Company and constructed a telescope. He also joined the Chicago Astronomical Society and learned how to use his telescope. After a short construction period, Dan began an observational program that concentrated on the bright planets. He joined ALPO in late 1978 and began to contribute his observations to that group. In 1979 he widened his astronomical program, joined the American Association of Variable Star Observers (AAVSO), and began to study variable stars. Everything went smoothly for Dan until December 17, 1979.

We have to stop here and drop back in time almost a hundred years. During the 1880s, the famous Mars observer Giovanni Schiaparelli, director of the Brera Observatory in Milan, Italy, noted that the north polar cap of Mars seemed to be divided into two parts. He determined that the cause of the division was a dark rift in the polar cap. He called this rift Rima Tenuis.

The rift was repeatedly observed during the next close approaches of Mars until the approach of 1918. The Rima Tenuis had disappeared. It was searched for repeatedly during following years but could not be seen. During the 1960s, Rima Tenuis was searched for using telescopes from 24 to 82 inches in diameter but these large instruments found nothing. When the Mariner—and later Viking—spacecraft arrived at the planet, their photography showed nothing at the north polar cap but layers of white. Rima Tenuis had indeed vanished.

On December 17, 1979, Dan was at his 10-inch telescope making drawings of Mars to forward to ALPO for study. He noted a dark rift that seemed to cut across the north polar cap and recorded it in his drawing. At the time Dan did not think much about the significance of what he was seeing. "I just tried to get it down accurately in the drawing," he said.

When Dan's drawing was received by ALPO's Mars director a few days later, it was looked at critically and with some disbelief. "I guess they didn't trust my eyesight," Dan said with a shrug. Astronomers around the world in ALPO's network of Mars observers rushed to their telescopes to see if Dan had really seen something that had not been visible on the red planet for almost 60 years. But for six weeks they had to wait. Because of the way Mars rotates, features visible near the edge of the planet pass behind Mars and are out of sight for that long. Finally, in February 1980, confirmations began to arrive: Rima Tenuis was back. Dan Troiani had succeeded with a 10-inch telescope where instruments two to eight times larger had failed. He had seen more from his backyard in Chicago than advanced spacecraft had seen from an orbit a few hundred miles above the planet. The credit for the rediscovery of Rima Tenuis on the north polar cap belongs to Dan Troiani, amateur astronomer, and no one else.

ANOTHER EXPERIENCE

Some professionals started out as amateurs, too—or just people interested in recording data on such objects as variable stars or comets. Take for example, Alan Hale, the co-discoverer with amateur astronomer Thomas Bopp of the famous 1997 comet Hale-Bopp. Hale didn't start out as a professional—he started as an amateur astronomer.

Hale started in astronomy like many of us—when he was young. "When I was in first grade by father checked out some books on astronomy from the local library and handed them to me to look at," Hale said. "I've never really looked back since, although my interests fluctuated a bit while in elementary school. I settled on astronomy 'for good' when I was in sixth grade. For a while I used my father's spotting scope, but shortly before I turned twelve, I managed to convince him to purchase a four-and-a-half-inch reflector from Sears (on sale, of course). And I never looked back."

There were other influences, too: (1) he was raised in a fairly small town in New Mexico, with lots of clear nights and no streetlights to speak of; (2) he watched the early space efforts (he vividly remembers watching some of the Gemini flights and the subsequent Apollo missions); and (3) he liked the television program *Star Trek*, with its premise of "exploring strange new worlds. . . ." And he kept reading every book on astronomy he could get his hands on. After a while, he decided he wanted to have a look at some of these objects he kept reading about.

His "evolution" from amateur to professional wasn't overnight. "I was involved in a science fair while in high school—and won first prize statewide in my division my junior year (the project was on the near-Earth asteroid Eros) and alternate to International Science Fair my senior year (the project was on Comet West)." Then he majored in physics at the Naval Academy and started the United States Naval Academy (USNA) astronomy club; spent 2 1/2 years working with the Deep Space Network at the Jet Propulsion Laboratory, and was involved with the VEGA Venus Balloon project and the Voyager 2 encounter with Uranus. He was also southwestern U.S. visual observations recorder for the International Halley Watch during this time.

He eventually returned to New Mexico and entered grad school at New Mexico State University (NMSU). "Although my dissertation work was on the subject of planets around other stars, and I've managed to publish a few papers on this subject," said Hale, "I continued 'backyard observing' after my return to New Mexico and still do to this day. It seems to be where I have achieved my biggest successes." Today, he is the director of the Southwest Institute for Space Research (home page: *http://www.swisr. org*).

Hale has some advice to those amateurs who want to make a discovery. "If ever you see a suspicious object, or something that doesn't look quite right, or something you don't recall ever seeing before, take a few minutes and check it out," suggests Hale. "Ninety-nine percent of the time it'll just be something that's always been there—and you've just never noticed it. But it's that other one percent that brings the pay-off. I learned my lesson in high school with Nova Cygni 1975, and it paid off for me twenty years later."

How does a successful discovery happen? Hale knows how: From Cloudcroft, New Mexico, Hale wheeled his telescope out into the driveway. It was a crystal clear sky, and he found the first comet he wanted to catch. But the other one was still close to the horizon. Hale fought the urge to go back indoors to wait until the comet rose higher. Instead, he searched toward a star cluster under the name M70—and saw a fuzzy object that had not been there before. "During the wee hours on Sunday morning, July 23, 1995, nature gave me a present," says the first line of Hale's book, *Everybody's Comet: A Layman's Guide to Comet Hale-Bopp*. What he and Bopp discovered independently was one of the brightest comets of the twentieth century.

2

THE EYE AND ASTRONOMY

*Fortunate is he whose introduction to the skies comes . . . through nature's
eyes alone and not through any telescope.*

LESLIE PETTIER

The basic instrument of the astronomer is the eye. Until Galileo Galilei came along in
1609 and turned his *optik tube* on the heavens, astronomy was done with the human
eye. The visual skills of early astronomers laid the groundwork that has taken us to
our present level of astronomical knowledge. Greek astronomers were able to explain
the phases of the Moon, devise a theory of planetary motions that Nicolaus Copernicus later revised and improved, measure the distance to the Sun and the Moon, and
determine the shape and size of the Earth. And all this without the use of a telescope.

Tycho Brahe, a Danish nobleman and astronomer, compiled accurate observations of planetary positions using some simple sighting instruments. After Brahe's
death, his assistant Johannes Kepler used this mass of data as the basis for his three
laws of motion. Without the contributions of these diligent observers, scientists such
as Sir Isaac Newton would have had no scientific shoulders to stand on-and the history of science might have been very different.

All of this work was accomplished using nothing but the eye. Anyone can apply
the same techniques and yield the same results. Yet one of the hardest things for the
beginning or casual observer is how to actually see and how this relates to astronomy.
Whether you observe with the naked eye, a pair of 10 × 50 binoculars or an 8-inch
telescope, everything you see has to pass through the eye.

HOW THE EYE WORKS

When you look at a star, its image enters the eye by passing through the *cornea* (Fig.
2-1), where the image is focused. The brightness of the image is controlled by the

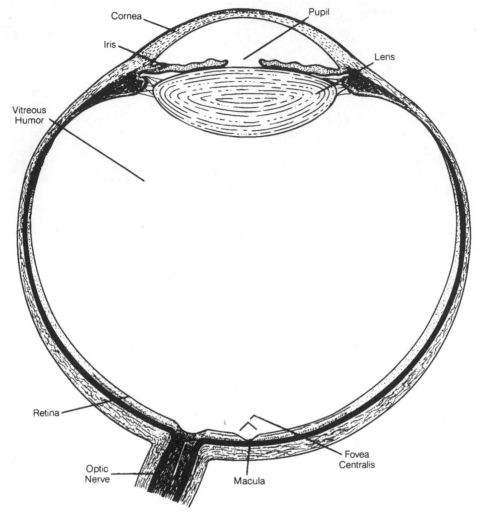

Cornea

Pupil

Iris

Lens

Vitreous
Humor

Retina

Optic
Nerve

Macula

Fovea
Centralis

FIGURE 2-1 *Cutaway of the human eye. (Courtesy Linda Warren, ©1988)*

varying diameter of the *pupil*, which opens or closes depending on the amount of light entering it. When you look at a bright object like the Moon, your pupils will *contract*, or close, because a large amount of light is reaching the eye (Fig. 2-2). When you look at a very faint object, like a galaxy or a very faint variable, your pupils will *dilate*, or open, to allow more light to reach the eye. The average size of the exposed pupil varies throughout life, but averages between 6 and 7 millimeters (mm). At 20 years of age, a healthy pupil can be as large as 8 mm when fully opened in low light. As a person ages, the effectiveness of the pupil diminishes. For example, at age 50, the pupil only opens to 5 mm, and by age 80, the pupil can only expand to a diameter of 2 mm.

Having passed through the cornea and the pupil, the image is given a fine focus by the *lens*. After moving through the central globe of the eye, the image falls on the *retina*, where the real power of the eye is located.

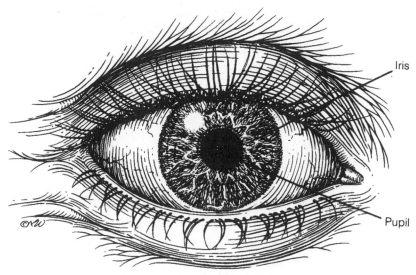

Iris

Pupil

FIGURE 2-2   *The pupil and iris act together to admit the light into the eye. The diameter of the fully open pupil is between 6 and 7 mm. (Courtesy Linda Warren, ©1988)*

The retina is a thin layer of cells lining the back of the eye. It is the only part of the central nervous system that is exposed to the outside world and direct stimuli. The retina is composed of five types of cells found in four layers and is able to change its sensitivity level. By adjusting to incoming visual signals, it is able, in effect, to be selective in what it passes on to the brain. Using its two outer cell layers, the retina can turn parts of itself on and off as it analyzes the image.

Located behind these two layers are the *photoreceptors*. They are anchored to a layer of cells from which they grow and draw nourishment. Because of the way they are attached, the photoreceptors face away from the incoming image, which must pass through the receptors before triggering a response.

There are two kinds of photoreceptors. The ones that allow us to see color are called *cones*, with an individual cone sensitive to one of three primary colors: blue, green, or yellow. The mixture of the signals from these three primary color receptors gives us color vision. The concentration of cones moves outward from an area of the retina called the *fovea centralis* (Fig. 2-3), which contains only cones, through the central part of the retina, called the *macula*.

A few degrees out from the fovea centralis the cones lose their dominance to the second kind of photoreceptor, called *rods*. Outnumbering the cones 20 to 1, rods are more sensitive to low light levels than cones. The rods play no role in color vision. Their primary purpose is to act as light collectors. Except within the central area of the retina, rods dominate. The tremendous number of rods allows the retina to gather enough light to see under almost any circumstance.

Because of the range of light intensity with which the eye must function, it has developed a dual system for vision. One system designed to see fine detail and detect rapid movement is called *photopic vision*, which makes use of the 6 million cones concentrated in the central area of the eye. In the fovea centralis, vision is the sharpest and the eye has its highest resolving power. For this system to work, however, light

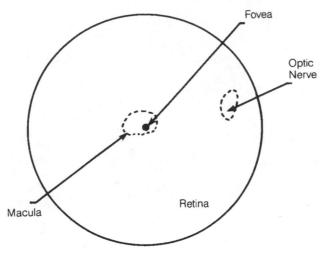

FIGURE 2-3 *The cones, the photoreceptors responsible for our color vision, are grouped in the fovea centralis. Surrounding the fovea are the millions of rods that enable us to see low light.*

levels must be extremely high. This is the system of vision that allows you to read the words printed on this page.

The other system, *scotopic vision*, functions when the light is dim, such as when you are looking through a telescope at a faint galaxy. Low light requires the eye to use the 130 million rods found everywhere in the eye except in the fovea centralis. These rods gather light from a large area over a longer period of time. You may be able to see the white of this page if you flip off the light. As your eyes adapt to the dark, you may even be able to distinguish the letters on the page as so many smudges—but you will not be able to read the words using this system of vision.

DARK ADAPTATION

The eye's ability to adapt to low light levels is very important to the observer. With the exception of a few objects such as the Moon and some of the planets, most light levels encountered by the amateur astronomer are very low. Adequate dark adaptation allows you to see details in galaxies and nebulae that the unadapted eye would let slip by unobserved.

When you expose your eye to low light levels, a number of physical changes take place. First, the pupil opens to allow more light to enter and strike the retina. Despite the greater amount of light entering the eye, only about half of it actually reaches the retina. The rest is absorbed by the fluids in the eye.

A chemical produced by the retina called *rhodopsin*, or visual purple, bathes the rods and helps the cells make the most of the small amount of light now reaching them. The chemically charged rods interact with the other cells to enhance the image before it is passed on to the brain. You can see this happen by sitting in a darkened room. At first you can see little, if anything. The longer you remain in the low light, however, the more details around you become apparent.

You need to encourage those same changes before you can see minute detail with the telescope. The entire dark adaptation process takes about 30 minutes. After the first minute of dark adaptation, the eye is 10 times more sensitive to light; after 20 minutes it is 6,000 times more sensitive. And after 30 minutes, the eye reaches its limit, with the fully adapted eye now requiring 30,000 times less light than normal to activate the retina. Although it is not necessary for you to sit in a darkened room, you can pass the time needed to adapt your eyes by sitting out under the sky and studying the constellations.

William Herschel, considered to be the founder of observational astronomy, went to extremes to obtain what he called "tranquillity of the retina." He would spend 20 to 30 minutes in complete darkness before he would try to observe faint objects. His dark adaptation was so complete that he would be forced to stop observing if a third-magnitude star came into his field of view. The light would almost blind him.

THE LIMITS OF VISION

The ability of any optical instrument is measured against the eye. The limits of the eye act as a baseline for comparison. It is important to understand those limits so we can learn how to make astronomical instruments that exceed them.

The astronomer has to contend with the fact that the eye can take in only so much light. The light-gathering power of any optical system is determined by the size of the area able to receive light. In the eye, this is the diameter of the fully dilated pupil. In a telescope, it is the diameter of the lens or mirror. The amount of light the eye, or optical surface, is able to gather determines how faint an object it can detect.

When you look at the stars, the first thing you notice is that they are not all the same brightness. Some are plainly visible even through the light pollution of the city; others are only visible well away from city lights. The brightness or dimness of stars and other celestial objects is measured in *magnitude*, a system devised by the Greek astronomer Hipparchus that has been virtually unchanged since 150 B.C. Sirius, Castor, Pollux, Vega, Deneb, and the other bright stars were lumped together as first-magnitude stars. Slightly fainter stars became the second magnitude. Hipparchus continued to categorize all the stars he could see, with the final group in his system, stars of the sixth magnitude, representing the limit of human vision.

Science has improved slightly on Hipparchus' system. Complex photometric devices instead of the naked eye now measure the brightness of the stars. The range of the magnitude scale has been vastly extended. Today, the dimmest star visible to the largest telescope is about twenty-fifth magnitude. But the basic ratio of brightness remains unchanged: Each magnitude is 2.5 times brighter than the next magnitude. A first-magnitude star is 2.5 times brighter than a second-magnitude star and 6.25 times (2.5×2.5) brighter than a third-magnitude star. Sixth-magnitude stars are 100 times fainter than stars of the first magnitude.

However, remember that the limit of human vision is dependent on a number of outside factors. Your *limiting magnitude*, the faintest star you can see, will be affected by your observing site. In the city, it is difficult to see down to the sixth magnitude; there are times when you can barely make out first-magnitude stars. On most nights in a big city like Chicago, you can usually see to fourth magnitude.

Someone with exceptional eyesight may be able to see to seventh magnitude under clear country skies.

There is also the matter of *resolution*. If you hold this book 12 inches from your eyes, you can plainly see each letter on the page. The eye is able to separate each letter into a distinct and recognizable shape; in other words, it can resolve the letters. If you move the book back 5 feet, can you still distinguish the letters? How about at 10 feet? At 20 feet? Somewhere down the line the eye stops seeing individual letters and they all run together. At that point, the eye has reached the limit of its resolution.

In astronomy, size and distance are measured in degrees. Each degree is made up of 60 arcminutes, and each arcminute is further divided into 60 arcseconds. *Resolving power* is measured with these smaller units. The generally accepted limit for the unaided eye is about 4 arcminutes, or 240 arcseconds. You can easily see the full disk of the Moon because it is 1/2 degree, or 30 arcminutes, in diameter. The star at the bend in the handle of the Big Dipper, Mizar, has a companion, Alcor, at a distance of about 12 arcminutes. (The ability to see Mizar and Alcor was a test of visual acuity in many cultures.) The double star Epsilon Lyrae is also a good test of visual acuity. With its components 208 arcseconds apart, it is just below the threshold of the unaided eye. A person with very good vision should be just able to separate the pair.

LIGHT POLLUTION

The ability of your eyes to see anything in the night sky is contingent on one thing: that you can actually see stars. Amateur astronomers in the major cities of the United States and around the world are faced with a growing menace called light pollution. Streetlights installed to curtail crime and make the streets safe have created a nightmare for many urban observers. Even amateurs living in rural areas are not unaffected by this growing problem. Ten years ago an amateur living 20 miles outside a major city like Chicago had a beautiful view of the sky. Observing is now tainted with the spreading glow of the city's lights, to say nothing of the lighting "improvements" installed in the suburbs (Fig. 2-4).

Light pollution problems fall into one of the three general categories. *Light trespass* is simply light spilling from one location onto another. If your neighbor has a security light on all night and it interferes with your observing, you are the victim of light trespass. Light trespass deals a death blow to the dark adaptation you have worked so hard to achieve and eliminates many of the faint, fuzzy objects inhabiting the night sky. The best way to deal with this is to diplomatically ask your neighbor to turn off the light. Offer him a look through your telescope—and in 99.99 percent of the cases he will be happy to oblige. If the offending light is a streetlight placed by the city, it is not a lost cause. Try to position yourself so that the offending light is blocked by a building or trees, or erect some kind of shelter to block the light.

Working hand in hand with light trespass is *glare*. We have come to expect a good light to be both dazzling and omnipresent. A good light, however, should not blind you when you look toward it.

The final category is *urban sky glow*. This has led to major problems for large observatories such as Mount Palomar in California, the Kitt Peak National Observa-

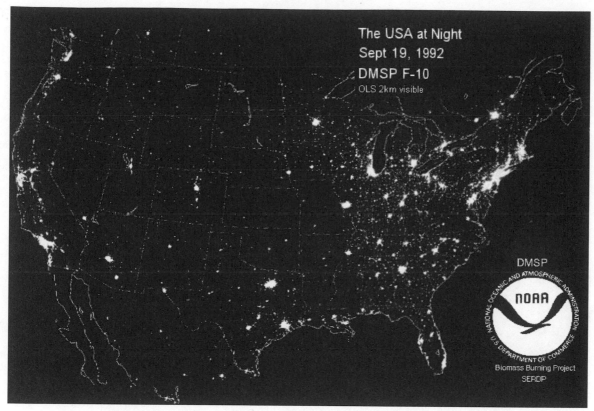

The USA at Night
Sept 19, 1992
DMSP F-10
OLS 2km visible

DMSP

NOAA

Biomass Burning Project
SERDP

FIGURE 2-4 *The United States at night from a satellite—and the abundance of light that contributes to light pollution. (Courtesy of the International Dark-Sky Association)*

tory in Arizona, and the U.S. Naval Observatory in Washington, DC. The thought of having to put such magnificent instruments in dark storage because of the encroaching lights of urban progress has rallied many professional astronomers into action (Figs. 2-5 and 2-6).

The obvious culprits in this battle against light pollution are lights. Now don't get us wrong-professional and amateur astronomers are not against nighttime lighting. Like everything else, however, there is a right way and a wrong way to illuminate the night. It makes no sense to use lights that waste more than 30 percent of their effectiveness by spilling light up into the sky (Fig. 2-7). Light that goes straight up is not useful in cutting down crime or protecting the citizens of the city, it is simply wasted. "About the only useful purpose for lights like this," says astronomer Dr. David Crawford of the Kitt Peak National Observatory, "is to light up the underbelly of airplanes." It is simple to install lights and fixtures that put light where it is most needed—on the ground.

Most cities and industries today use what is called high-pressure sodium (HPS) light. This type of light is not energy efficient, nor are the fixtures in which these lights are mounted. A better solution would be to install a more economical type of light and shield it so that the excess light does not seep into the sky. This type of light—called low-pressure sodium (LPS) lighting—does exist. When used in con-

FIGURE 2-5 *The light from Los Angeles in 1908. (Courtesy of the International Dark-Sky Association)*

junction with properly shielded lighting fixtures, it reduces light pollution and improves the overall economy of the lighting system.

The light emitted by LPS fixtures is restricted to a narrow band of yellow light that is easily filtered—another plus for this type of lighting. LPS lights are easy to recognize by their yellow glow. HPS lights, identified by their pinkish color, put out their light over a much broader spectrum and are almost impossible to filter out.

And there have been many successes. Agreements between astronomers and the city of Tucson, Arizona, have lead to the installation of the more efficient LPS lighting system. This saves the city money and allows the nearby observatory to filter the light for more efficient viewing. It is a solution that lets everyone win.

There are other examples. There has been a great deal of progress in outdoor lighting control made in communities on the southwest coast of Florida. This area is known for its rapid expansion, and it has gone from an area of trailer parks and mobile homes to zoned areas and regulated land development—with some calling it one of the most environmentally aware regions of the world. This also includes local awareness of light pollution and strict lighting controls—not only for astronomical reasons, but also for ecological reasons such as migratory birds, the hatchling sea turtles' marches to the sea, and sundry other wildlife. For example, four sections of the zoning statutes of Sanibel Island deal with outdoor lighting—all limiting the use of light and cutting back light pollution.

FIGURE 2-6 *The light from Los Angeles in 1988, showing the increase in light pollution. Compare this image with Fig. 2-5. (Courtesy of the International Dark-Sky Association)*

Another area that won a fight against potential light pollution was the area around Mt. Hopkins in southern Arizona, the site of the Whipple Observatory and the Smithsonian Astrophysical Observatory. An Arizona developer wanted to put in a 6,100-home development site at the base of the mountain—and even with strict light pollution controls, would still have increased light pollution by 8 to 14 percent. But there was an immediate cry—not only from the astronomers, but through an online petition started on the Internet from people all around the world. With a vote of 4 to 1 from the Pima County Board of Supervisors, the development was defeated—and even more stringent outdoor lighting control ordinances are now being worked on for the area.

What can the amateur astronomer do about light pollution? The entire issue comes down to the simple task of education. Many people outside of astronomy don't know the first thing about light pollution—and you, as an affected observer, can help introduce them to the problems and solutions concerning light pollution. You can show people, especially neighbors who have those wonderful security lights, how to improve the lighting so it is more effective and efficient. If you are successful, your neighbor can get a lower energy bill and you can get a darker observing site. Again, everybody wins.

The International Dark-Sky Association (IDA) has been organized for the purpose of preserving dark skies for astronomy. Dr. David Crawford, one of IDA's

FIGURE 2-7 *The glare from a billboard along the road. (Courtesy of the International Dark-Sky Association)*

founders, estimates that in a few generations "the majority of people in the world will never have seen the universe. The only place they will be able to see it is in a planetarium or on a TV screen. And that," he adds, "is sick."

NAKED-EYE ASTRONOMY

Probably the most enjoyable way to discover the universe is with the eye alone. Many objects and phenomena can be observed using nothing more than a comfortable chair and your eyes.

The first thing to do if you are going to enter the wonderful world of amateur astronomy is to learn the constellations. Now this may sound like a real task but it is worth the effort. Learning the shapes and stories behind star patterns moving through the night sky can become an avocation in itself.

Many people complain that they just can't see much from the city. Mike lived in Chicago all of his life, and he watched the skies gradually deteriorate. He remembered the skies of Chicago he saw as a boy—and if he looked hard enough, he could just make out the Milky Way. Then that changed. From where he observed only a short time ago—within a mile or so of Midway Airport and right under a landing path—he could barely make out fourth-magnitude stars.

Strange as it may seem, learning the sky under conditions like this can be a blessing in disguise. The sky from a dark site is so strewn with stars that it can confuse new observers. Such confusion can lead to a decision to find another outlet for scientific curiosity. Learning under the limited skies of a city like Chicago allows you to learn the brightest stars in each constellation, making the transition to darker skies and more stars easier.

At the end of this book, a bibliography mentions many constellation guides. We suggest that you visit your local library and study one or two of them. Take the books home and leaf through them at your leisure. Remember you will not be tested on any of this, so there is no pressure. Take the time to learn the constellations. If you never venture any further into the world of amateur astronomy, you will have a treasury of knowledge that no one can take away from you.

3

BINOCULAR ASTRONOMY

The stars . . . appear with exquisite and startling beauty through a field glass.

WILLIAM T. OLCOTT

The eye may be the basic instrument of the astronomer, but it needs a little help to get a close look at the universe.

Probably the easiest and fastest way for anyone to get involved in observational astronomy is to pick up a good pair of binoculars. For the beginning amateur trying to decide on the type of telescope best suited to his or her needs, a pair of binoculars offers a convenient way to become acquainted with the sky. After spending some time learning the constellations with the naked eye, good binoculars can show you the full extent of what lies just beyond.

The stereo view of binoculars makes the stars stand out in 3-D. Because you do not strain your eyes, you are a more relaxed observer. Besides, using binoculars is a natural way to bridge the gap between naked-eye observing and the use of a telescope of any size. Using binoculars allows the beginner to develop the skills necessary to get the most out of a telescope.

It isn't a bad idea for the more advanced amateur to pick up a pair of binoculars, too. After crouching at a telescope for any length of time you can forget the grand, sweeping vistas offered by binoculars. Using binoculars in conjunction with the telescope will also allow you to detect faint objects, then determine the best way to position the main telescope for a better look.

BINOCULAR FACTS

Binoculars come in all shapes and sizes (Figs. 3-1 and 3-2). The light path in a pair of binoculars does not run straight through the instrument as it does in a refractor tele-

FIGURE 3-1 *7 × 35 binoculars.*

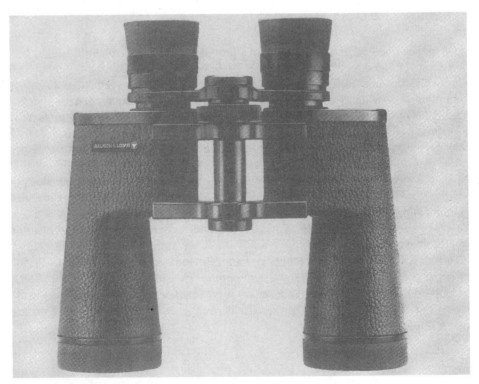

FIGURE 3-2 *10 × 40 binoculars.*

scope. If it did, the resulting image would be upside down and very hard to interpret. Binoculars make use of prisms that correct the orientation of the image and make the light path from the objective to the eyepiece short and easily managed. *Porro prism* binoculars use two side-by-side prisms to bend the light and correct the image. *Roof prism* binoculars use a more compact set of prisms to direct the light. Roof-prism binoculars are lighter and more compact than those using the more conventional porro prism (Fig. 3-3).

Binocular size is defined by two numbers separated by a × symbol. Examples of different size binoculars are: 7 × 35, 7 × 50, 10 × 50, and 20 × 80. The first number is the magnification of the instrument; the second is the size of the objective. The mag-

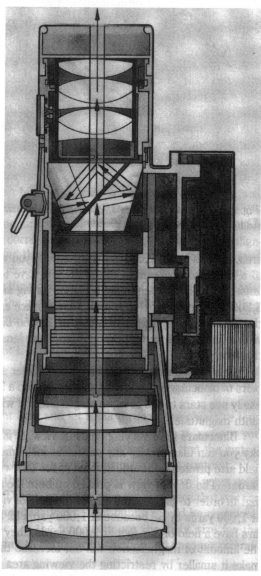

FIGURE 3-3 *The internal workings of one type of a binocular.*

nification is the number of times larger the image appears than if viewed by the naked eye. A 7 × 35 instrument will yield an image 7 times normal size; a 20 × 80 binocular will give you an image 20 times larger than that seen by the eye alone. One important thing to remember is that the higher the magnification of any optical instrument, the more difficult it is to keep them and the image steady.

Binoculars up to a magnification of 10 are fine to hold in your hand, but if you want more powerful binoculars you will need some type of mount. Many of the more powerful binoculars, such as 11 × 80 and 20 × 80, come with either a separate bracket or a built-in socket for mounting them on a photo tripod.

Binoculars enable the amateur astronomer to increase the light-gathering power of the eye. The most important factor in determining the amount of light taken in by an optical system is the size of the area that receives the light. In the eye this receiving area is the pupil; in the binoculars it is the objective lenses. If you want to determine the comparative light-gathering power of two lenses, you have to compare the two areas available to receive the light. If the size of a lens is doubled, its area increases by four times. In other words, the bigger the lens the more light it can gather.

If a lens is 25 mm in diameter, the surface area available to receive light is about 1,550 square mm. If you want to compare its light-gathering power to a lens that is 50 mm in diameter, you have to compare the surface areas of the two lenses. The 50-mm lens may be twice as big as the 25-mm lens, but its surface area is 6,200 square mm or four times the surface area of the 25-mm lens. So if we compare the light-gathering ability of a pair of 10 × 50 binoculars to that of an average observer with a 7-mm pupil, we find that the binocular lens has a surface area over 50 times larger than the pupil.

This means that you can see much fainter stars when you use binoculars. For example, two pairs of binoculars to help with observing are a 7 × 35 and a 10 × 50. With the 7 × 35, you can easily see stars to the seventh magnitude; with the 10 × 50 instrument, an observer can see ninth-magnitude stars.

Binoculars have varying fields of view. It is important to know how large an area of sky you can take in with any particular instrument. All binoculars have an expression of field size printed on them. In most cases this will be a statement like "357 feet at 1,000 yards." The field of view is printed on both the instruction book and the instrument casing. In order to find out the size of the field of view in degrees, divide the number of feet at 1,000 yards by 52 (because at 1,000 yards 52 feet equal one degree). So if your binoculars have a field of 357 feet at 1,000 yards, they have a 7-degree field of view. Increasing the amount of magnification will not increase the size of the field of view. In fact it will make it smaller by restricting the viewing area of the instrument.

When you are observing faint objects in the night sky, it is important to have as bright an image as possible. The objective lenses of the binoculars contribute most of the power required to create a bright image, but you still need a little help from the binocular eyepieces.

This is where the *exit pupil* becomes important. This is created by the eyepiece and in drawings looks like a little cone of light extending away from the eyepiece. The size of the exit pupil is important to the brightness of the image that you see. As we mentioned before, most people have a normal dilated pupil size of 7 mm. For the brightest image, the exit pupil should be as large or larger than the pupil size so that the light will fill the entire pupil. The exit pupil can be easily figured for any type of binoculars by dividing the objective size by the magnification. A pair of 7×35s have an exit pupil of 5, as do a pair of 10×50s.

SELECTING AND TESTING BINOCULARS

Selecting a pair of binoculars is not as easy as one might think. Binoculars can serve a multitude of functions. In addition to being used to observe the stars, they can go to sporting events or on nature field trips. If you intend to use them for such things it is best to get a general-purpose pair such as 7×35s, 7×50s, or 10×50s. Many manufacturers are now making large objective instruments: 11×80s, 20×80s, and even 25×120s. These are special types of instruments and are really only suited for astronomical use. Regardless of the size you decide on, remember to get a pair with a large field of view and an adequate exit pupil.

There are a number of steps you can take when purchasing a pair of binoculars to make sure you are getting a quality instrument. First, check the alignment of the binoculars. The barrels, or the sections you look through, should point in the same direction. If they don't, you will see a double image for a brief instant while your eyes correct the problem. If you suspect poor alignment, focus the binoculars on an object and then draw them slowly away from your eyes to a distance of a few inches. The image should remain single. Next, with the binoculars still focused, slowly blink your eyes a few times. If the pair is out of alignment, it will take your eyes a couple of seconds to realize that the images are not properly aligned before they correct for the misalignment. Using a pair of binoculars with this defect will result in eyestrain and a headache.

The barrels should flex freely but not loosely as you move them together. Give the barrels a slight twist. There should be no play in the hinge between them. The focusing mechanism should work smoothly. The eyepiece frame should remain stable and not wobble when you are racking the eyepieces in and out.

Now stand with your back to a bright light and let it fall through the objective lens onto the inside of the barrel. The interior surfaces should be clean. The reflections you see on all the optical surfaces should be purple or amber. This indicates that the instrument has antireflective coatings that will increase light transmission and contrast.

Turn the binoculars around and hold them up to a bright surface. You will see a small disk of light "floating" in each eyepiece. These are the exit pupils. It is here that you can best tell if the manufacturer is skimping on quality. The exit pupils should be complete, round disks. If they are, the prisms used in the binoculars are the right size for the instrument. Many manufacturers are trying to get by with smaller and cheaper prisms. These inferior prisms give the exit pupil a squared-off look because they cut off some of the light. The BAK-4 prisms give excellent exit pupil definition.

But there are good prisms. For example, images seen through the instrument should be bright and sharp. Gray or filmy images indicate that there is a problem with the contrast in the lens set. The field of view should be wide and bright. Check the quality of the field edge by moving the instrument over a straight line, such as a roof edge or a telephone line, at a right angle. Notice if the line tends to distort as it moves into and out of the field of view. Take a look at a line dividing a bright area from a dark area. Because no lens surface is totally free of chromatic aberration, you will see some red and blue along the line. The less you see of these colors the better the instrument.

Finally, look at a brick wall. The image of the bricks should be equally focused across the field of view. In an inferior pair of binoculars the center is sharply focused but the edges may appear bowed and curved. This indicates that the binoculars' field of view is not flat. Using them over an extended period of time will lead to fatigue.

It is important to get a quality pair of binoculars, but you cannot judge quality solely on price. Prices may vary greatly. A pair of 7 × 35 binoculars may range in price from $24 to hundreds of dollars. The manufacturers of some expensive binoculars take short-cuts in production that would surprise you, while some seemingly cheap instruments are sturdily built with good optics. When you are deciding on a pair of binoculars, look for ones that have the best optics and largest aperture—and still fit your budget.

Regardless of price or size, remember it is how the instrument is used that is important. Great discoveries are not made by instruments, but by the observers using those instruments.

OBSERVING WITH BINOCULARS

The first thing you need to remember about using binoculars for astronomical observation is comfort. Standing with your neck craned, holding binoculars over a long period of time can become very tiring. The best way to observe is from a reclining position. A reclining yard chair with arms gives you a relatively stable platform from which to observe. It will also allow you to keep an eye on the sky with less strain to your neck and arms.

If you have to stand, it is best to mount your binoculars on a photo tripod or some other kind of stand. You can even take binocular observation a step further and build an observing stand that allows you to look down with no strain. Many stands are constructed using a simple first-surface, flat mirror that reflects the light from the stars up into the binoculars' objectives. Using a stand like this allows you to relax and observe from a seated position.

A good object to look at through your binoculars is the Moon. With good 7 × 50 or 10 × 50 binoculars, you can follow the line of sunrise as it crosses the lunar surface. The *terminator* throws the Moon's features into sharp relief, allowing you to study most of the larger craters.

We've talked about the advantages of using binoculars to observe, but what can you see with them? Good binoculars will show you all the planets of the solar system with the exception of faint Pluto. If you have optically excellent binoculars you will be able to make out the phases of Venus as it moves around the Sun. The four bright satellites of Jupiter are easily seen. If you have a steady hand or a stable mount you may be able to see the rings of Saturn. With a better pair of binoculars, you may even make out the dark smudges on Mars. Mercury, Mars, Uranus, and Neptune, unfortunately, will only show up as points of light because these planets are either very small or far away.

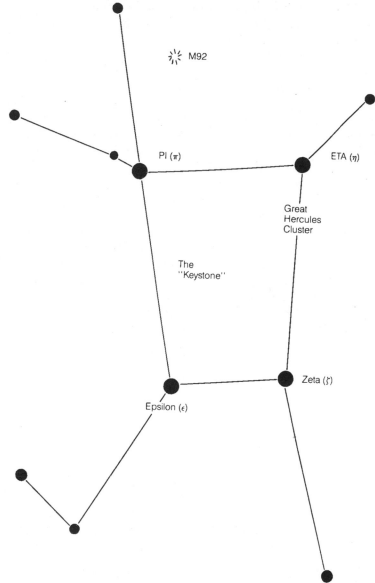

M92

PI (π)

ETA (η)

Great
Hercules
Cluster

The
"Keystone"

Zeta (ζ)

Epsilon (ε)

FIGURE 3-4 *Location of the Great Hercules cluster.*

The best thing for you to do is begin by studying the constellations using the greater light grasp of your binoculars. If you live in the city or are bothered by severe light pollution, the additional universe that a good pair of binoculars puts at your disposal will astound you. After taking some time to reacquaint yourself with the new views of the constellations that your binoculars give you, it is time to meet a few of the denizens that inhabit the great bowl that lies above our heads.

In early July, Lyra is almost directly overhead, so it is a hard constellation to miss. The brightest star of the group is Vega, the fifth brightest star in the sky.

Trailing south from Vega is a grouping of two third- and two fourth-magnitude stars in the form of a parallelogram. These make up the main body of the constellation. Turn your binoculars on the star at the northeastern corner of this figure and you will find yourself looking at Delta (δ) Lyrae. You will see two stars, one red and the other blue. Delta is what is known as a *double star*-two stars held together by gravity and revolving around a common point in space.

Directly, opposite Delta Lyrae is Beta (β) Lyrae. If you watch Beta over a period of about 12 days you will notice that it is not constant in brightness. Beta is a *variable star*. It varies in brightness from magnitude 3.4 to magnitude 4.1. Compare Beta to the star at the end of the parallelogram, Gamma (γ) Lyrae. Gamma shines at magnitude 3.25. When Beta is at its *maximum*, or brightest, it is almost as bright as Gamma. At its dimmest, or *minimum*, it shines at only 65 percent of Gamma's brightness. Beta's variability is caused by a companion star that moves around it. Every 12 days Beta is eclipsed by its companion, causing the light we see to dim.

Moving directly west from Lyra, you will come across another faint constellation, Hercules (Fig. 3-4). Four stars, three of third- and one of fourth-magnitude, make up the identifying keystone asterism, or star grouping, of the constellation. Our next target is located between two of those third-magnitude stars, Eta (ε) and Zeta (ζ). Between them you will find probably the best example of a *globular cluster*-the Great Hercules cluster. In 10 × 50 binoculars, the cluster appears as a soft glow surrounding a bright center (Fig. 3-5). It is a grouping of hundreds of thousands of

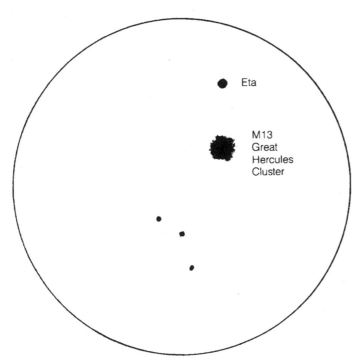

FIGURE 3-5 *The Great Hercules cluster, M13, as seen through 10 × 50 binoculars—a "smudge" of light against the sky.*

FIGURE 3-6   *The Great Cluster, M13, through a telescope. (Courtesy R. M. Sandy and NASA)*

stars packed together so tightly that its center cannot be resolved into individual stars even by the largest telescopes (Fig. 3-6).

Under the skies of a big city like Chicago, Vulpecula is a barely visible constellation just south of Cygnus. It passes overhead in early September. Turning your binoculars on the area shows you a wealth of detail invisible to the naked eye, even under the best of skies. The Milky Way is rich in the area. What was formerly an area marked by two fourth-magnitude stars becomes choked with the brightness and diversity of our galaxy.

Amid this multitude of stars, the very fine *planetary nebula* known as the Dumbbell nebula becomes visible. Nebulae like this are the result of a star blowing its outer surface into space. The gases and matter expelled form a shell that slowly moves out from the star into space. In most cases the star that suffered the explosion is invisible to all but the largest telescopes. All we get to see is the beauty of the aftermath. With the added contrast of the binoculars producing a darker sky, the nebula floats ghostlike in the field.

Another of our denizens of the deep sky is found in the autumn constellation of Andromeda. The main asterism of the constellation is made up of two arcs of stars. The eastern arc has one third- and two second-magnitude stars and the smaller, fainter, more westerly arc made up of three fourth-magnitude stars. Both arcs are linked by the second-magnitude star Alpha (α) Andromedae. To locate our next target, find the middle star of the eastern arc, Beta (β), and move west past the second arc. In your field you will see a faint, greenish glow. That is the Andromeda Galaxy. Similar to our own galaxy, the Milky Way, Andromeda is 2 million light years away from us. This may seem far, but it is the closest *spiral galaxy* to us.

A few hours after Andromeda passes overhead, the constellation Taurus becomes visible. With the red first-magnitude star Aldebaran gleaming and the Pleiades splendidly displayed just above the V of the Hyades, Taurus is easy to locate. The Hyades and Pleiades are two examples of *open clusters*. Because of the size of these magnificent star clusters the only way to take in all of their beauty is with a pair of binoculars. Both of these objects are groups of stars moving together through space.

Southeast of Taurus is probably the most recognizable constellation in all the sky. Orion, the hunter and eternal adversary of Taurus, holds many deep-sky treasures. Hanging from Orion's belt is his sword, and near the tip of this is a glow that is easily seen with the naked eye—the Orion nebula. If you turn your binoculars on this object, you will see what appears to be clouds of gas embedded with stars—and that is just what you are seeing. The Orion nebula is a *reflection nebula*. The colors and texture you see are lit by the stars that are glowing inside the nebula. Unlike a planetary nebula, which is created by a dying star, reflection nebulae are the birthplaces of stars.

These are the basic types of deep-sky objects that a pair of binoculars will reveal to you. Now that you have a taste of the vastness and beauty of the visible universe, it is time to take the next step in our journey. Let's look at some of the tools used by astronomers.

TOOLS OF THE ASTRONOMER

There is no greater relaxation than to embark on a celestial sight-seeing trip with a . . . telescope.

WILLIAM T. OLCOTT

The time will come when you will want to look deeper into the universe. For that you will need a telescope. It is the telescope that really allows the eye to see the wonders of the universe. But before you go out and plunk down your hard-earned money, you will need a few facts. It is hard to take advantage of an informed consumer.

A glance through any of the popular astronomy magazines yields a seemingly unending number of telescope shapes, sizes, and varieties. But when all the fancy trappings are stripped away there are only two basic types of telescopes-and they operate using the same principles (Fig. 4-1). Both collect light and produce an image where the collected light comes to a focus. The part of the telescope that collects light is called the *objective*. In the refractor telescope the objective is a lens; in the reflector telescope it is a mirror.

The distance it takes for the objective to form an image is called the *focal length* of the telescope. Directly related to the focal length is the *focal ratio* (f-number) of the telescope. The focal ratio of the instrument can be found by dividing the focal length by the objective diameter. These figures can be expressed in inches, millimeters, or centimeters as long as the same unit of measure is used throughout the process. Most manufacturers measure focal lengths using the metric scale.

A 6-inch objective with a focal length of 30 inches has a focal ratio of f/5. If it had a focal length of 48 inches its focal ratio would be f/8. The smaller the focal ratio is, the shorter the focal length and the faster the system. The larger the focal ratio is, the longer the focal length and the slower the system. A 6-inch f/5 is faster than a 6-inch f/8. A 6-inch f/10 is slower than both of them.

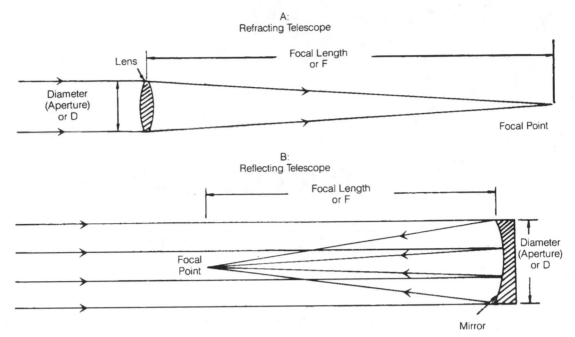

FIGURE 4-1 *Telescopes operate by bending light. A refracting telescope (a) uses a lens while a reflecting telescope (b) uses a mirror. (Courtesy Tasco)*

This speed designation is useful when you consider what you want to observe with your telescope. The faster the system is, the larger and brighter the field of view. Fast systems, say f/4 or f/5, are ideal for looking at very faint, diffuse objects such as star clusters, nebulae, and galaxies. Slower systems, f/10 and up, give better contrast in the field of view and excellent views of bright objects such as the Moon, planets, and double stars.

FUNCTIONS OF A TELESCOPE: LIGHT GRASP

The most important aspect of any astronomical instrument is its ability to act as a light collector. This is known as the telescope's light-gathering power or *light grasp* (binoculars also have a certain light-gathering power, but the term *light grasp* is used mainly in reference to telescopes). The light-gathering power of a telescope is a function of the size, or *aperture*, of the objective. The light-gathering power of the objective is compared to that of the fully dilated pupil of a healthy eye, which has an aperture of 6 or 7 mm.

As we have seen with binocular objectives, the light-gathering power of the telescope objective increases as its size increases. So the light-gathering power of a 4-inch objective is not twice, but four times, that of a 2-inch. Although the diameter of the objective has only doubled, the area of the objective has increased four times. The total area of the objective collects light. The larger the objective, the greater the light grasp of the instrument and the fainter the objects it can see. Table 4-1 shows you the comparative light-gathering power of the human eye and various size objectives.

TABLE 4-1. LIGHT-GATHERING POWER

APERTURE (MM)	APERTURE (IN.)	LIGHT-GATHERING POWER (EYE = 1)
35	1.4	25
50	2.0	50
60	2.4	80
50	3.1	130
100	4.0	210
150	6.0	480
200	8.0	840
250	10.0	1,300
317.5	12.5	1,900
330	13.1	2,100
355	14.0	2,400
405	16.0	3,360
444	17.5	3,680
500	20.0	4,900

LIMITING MAGNITUDES

The faintest star the unaided eye can see under perfect conditions is sixth magnitude. With the increased light grasp of even a small 50-mm telescope you may see stars almost 100 times fainter. But no matter how big a telescope is, it can only see stars of a certain brightness. In other words, all telescopes have a limiting magnitude, or the faintest star that can be seen through a telescope under ideal conditions. The limiting magnitudes of various size telescopes are listed in Table 4-2. These numbers do not, take into account such factors as the visual acuity or skills of the observer, however, and they have been exceeded by more than one observer. Nonetheless, a quick look at Table 4-2 would tell an amateur astronomer with a 60-mm refractor telescope that it would be futile to try to locate the planet Pluto. After all, the tiny planet is usually found around magnitude 14, well past the limit of the observer's instrument.

IMAGE BRIGHTNESS

Amateur astronomers observe two types of objects. Stars are *point source objects*. No matter how big a telescope you use they will always be seen as points of light. The Moon and most planets, on the other hand, show a definite disk when viewed through even a small telescope. Because of this they are called *extended objects*. Star clusters, nebulae, and galaxies are collectively known as *deep-sky objects*. They compose another class of extended objects.

The fact that an 80-mm refractor is capable of reaching magnitude 11.5 can be a bit deceiving. What it really means is that you may be able to see point sources, or stars, to that magnitude under ideal conditions. It does not always mean you can see deep-sky objects of that magnitude. The brightness of the image of a deep-sky object is not concentrated in one central point, it extends over the entire visible area of the

TABLE 4-2. LIMITING MAGNITUDES

APERTURE (MM)	APERTURE (IN.)	LIMITING MAGNITUDE
35	1.4	9.5
50	2.0	10.5
60	2.4	10.9
80	3.1	11.5
100	4.0	12.0
150	6.0	12.9
200	8.0	13.5
250	10.0	14.0
317.5	12.5	14.5
330	13.1	14.6
355	14.0	14.7
405	16.0	14.8
444	17.5	15.0
500	20.0	15.5

object. While it may have a high magnitude rating, its overall surface brightness can be low, making it hard to see.

An object like the spiral galaxy M33 is a good example of this phenomenon. Its visual magnitude is listed as 6.7, and if you are far enough away from the annoying lights of the city you can just make it out against a dark sky. It looks like a ghostly mist covering an area about the size of the full Moon. But the magnitude that is listed for the galaxy is what is called an *integrated magnitude*; that is, it is what the magnitude of the galaxy would be if it were seen as a point source. Because M33 covers such a large area of the sky its light is spread out. Therefore, if you look for it through your telescope it may be impossible to find.

FUNCTIONS OF A TELESCOPE: RESOLUTION

Diffraction is a phenomenon of light that allows us to see detail in the universe. Diffraction causes light to bend ever so slightly when it encounters an obstacle, such as when it enters a telescope tube. This bending causes the resultant image of the star to distort and appear as a minuscule, blurred dot through the telescope. This image of the star is called a spurious disk or *airy disk* (Fig. 4-2).

Named for the seventh astronomer royal, Sir George Airy, the airy disk looks like a tiny bull's-eye of light formed by a bright central disk surrounded by alternating dark and bright rings. The disk's greatest brightness is found in the center, and it drops off to almost nothing at the first dark ring, also called the first *interspace*. The brightness then increases again, although to only a fraction of the central brightness, for each discernible ring. With a 6-inch telescope and good seeing, a first-magnitude star will show as many as five or say of these diffraction rings.

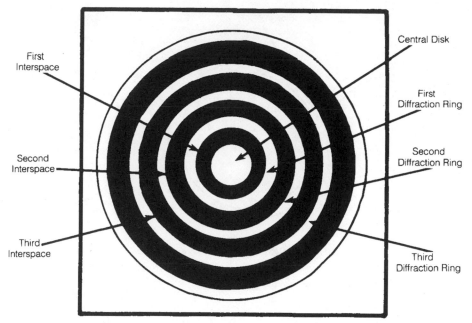

First Interspace

Central Disk

First Diffraction Ring

Second Interspace

Second Diffraction Ring

Third Interspace

Third Diffraction Ring

FIGURE 4-2   *The airy disk.*

THE RESOLVING POWER OF A TELESCOPE

When two equal point sources of light are close together, as with a double star, the airy disks overlap and the images are combined into a single airy disk (Fig. 4-3). When you are able to discern this disk into its individual components through the telescope, it is said to be resolved. The nineteenth-century English physicist Lord Rayleigh established the resolution with the formula: $R = 5.5/D$, where D is the aperture of the telescope in inches (5.5 is in arcseconds) and the resolution figure, R, is known as the *rayleigh limit* of the telescope. This occurs when the two images are separated by the first diffraction interspace.

William Dawes, a nineteenth-century English amateur astronomer, noticed that this theoretical limit could be exceeded by about 5 percent under good conditions. He further redefined the resolution limit of a telescope as $R' = 4.56/D$, again where D is the aperture of the telescope in inches, and R' (in seconds) is known as the *dawes limit*. This differs from the rayleigh limit in that it is the first point at which a double star is elongated enough to suspect the presence of two stars. For example, using this formula, we see that a 3-inch refractor can resolve two stars as close as 1.52 inches ($R' = 4.56/3$), and a 6-inch refractor can get down to 0.76 inches ($R' = 4.56/6$), but only if the two stars are about the same magnitude and you have great viewing conditions.

Similar to light grasp, resolution is dependent on the aperture of the instrument. The only way to increase the resolving power is to increase the size of the aperture. No amount of magnification will enable you to resolve a double star that is beyond your telescope's limit. Under actual observational conditions these limits are rarely reached because a number of factors—including atmospheric stability and the visual sensitivity of the observer—can combine to keep a telescope from reaching its resolution limit.

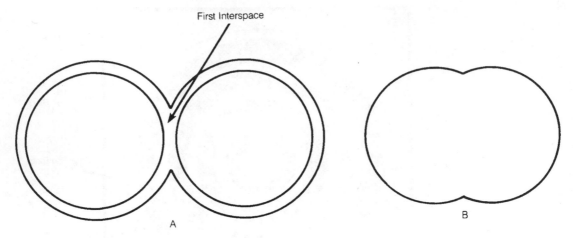

First Interspace

A

B

FIGURE 4-3 *A double star is said to be resolved at its traditional rayleigh limit (a) when the two airy disks are separated by the first interspace. The dawes limit (b) only requires them to be elongated enough to tell if there are two stars.*

The resolving power of a telescope is applied most often to double stars, but this is not the only area where it is important. The resolving power of a telescope dictates such things as the sharpness of detail visible on a planetary disk and the smallest crater visible on the Moon. Table 4-3 gives the rayleigh and dawes limits for various size telescopes.

TABLE 4-3. RESOLUTION LIMITS

		RESOLUTION IN ARCSECONDS	
APERTURE (MM)	APERTURE (IN.)	RAYLEIGH LIMIT	DAWES LIMIT
35	1.4	3.9	3.2
50	2.0	2.8	2.3
60	2.4	2.3	1.9
80	3.1	1.8	1.5
100	4.0	1.4	1.1
150	6.0	0.9	0.76
200	8.0	0.68	0.57
250	10.0	0.54	0.45
317.5	12.5	0.43	0.36
330	13.1	0.43	0.34
355	14.0	0.38	0.32
405	16.0	0.34	0.28
444	17.5	0.31	0.26
500	20.0	0.27	0.22

FUNCTIONS OF A TELESCOPE: MAGNIFICATION

The image formed at the primary focus of a telescope is slightly magnified. The eyepiece of the telescope acts on the image and increases it to a size large enough to be examined in detail. This process is commonly known as *magnification* and is the least important function of a telescope. Unlike the other functions of the telescope, magnification does not depend on the size of the aperture. Rather the focal length of the objective combined with the focal length of the eyepiece determines the instrument's magnification power.

To find the magnification of an eyepiece divide the focal length of the telescope by the focal length of the eyepiece. The result is the magnification power of the eyepiece. A 3-inch refractor telescope with a focal length of 910 mm and a 10-mm eyepiece has a magnifying power of 91 times or 91×.

To increase the magnification you have to change the focal length of either the objective or the eyepiece. Changing the focal length of the objective would mean replacing the objective, because each telescope comes with an unalterable focal length. It is easier to replace the eyepiece with one of a different focal length. This has its own problems, however. As the magnification increases, the focal length of the eyepiece becomes smaller and the eye must get closer to the eyepiece, until it is touching the lens. This is both uncomfortable and a waste of light.

It is much more important to consider the size of the exit pupil of the eyepiece. The exit pupil, also called the *ramsden disk* in relation to telescopes, is a small circle of light that seems to float above the eyepiece. It is the image of the telescope's objective illuminated by the object at which you are looking. It should be just as large as the pupil size of your eye. This allows it to transmit all the light gathered by the telescope to your eye so you can see the brightest image. On the other hand, if you are observing a very bright object such as the Moon, your pupil will contract and reduce the amount of light falling on the retina because the extra light isn't needed to form a sharp image. So a smaller exit pupil will suffice to fill the smaller size of your eye.

The exit pupil imposes a range of magnification on a telescope, and it is best if the user stays within that range. If the exit pupil of the eyepiece is so large that the eye's pupil cannot take in all the light, the telescope is working at a disadvantage. The full light-gathering power of the objective is not being used. In effect, you are decreasing the size of the telescope.

It is generally accepted that the lower limit of magnification is three or four times the diameter of the objective in inches, expressed as $3D$ or $4D$. A limit lower than this would rob the telescope of its ability to resolve an image sufficiently. Using this formula, a 3-inch refractor telescope would have a lowest useful magnification limit of 9 to 12 power. If the focal ratio of the instrument is f/11, the desired exit pupil is between 6 and 7 mm.

The highest allowable magnification is generally accepted to be 25 times the aperture of the objective, or $25D$. In our example, this would give a magnification of 75 for a 3-inch telescope with an exit pupil of 1 mm. Using too much magnification is self-defeating. After a certain point, the only thing magnification will do is reduce the sharpness of the image.

All this talk of exit pupils and magnification ranges should not scare you. Like anything else, there are exceptions to the rule such as specific observation situations. Certain objects take magnification better than others. To separate double stars, provided they are above the instrument's resolution threshold, you sometimes need the maximum magnification you can get from your instrument. Observing the planets, with their wealth of detail, also calls for higher magnification to make the most of the telescope's resolution ability. In some instances magnifications as high as 50 or 60 times the objective can be used, but those circumstances depend on steady eyesight (and atmosphere) and the type of object you are observing. Double stars and planets can take a lot of magnification.

OPTICAL IMPERFECTIONS

No optical instrument is perfect. This is as true for the telescope as for the eye. All telescopes are subject to five types of imperfections called *aberrations*. No matter how much you pay for an instrument, or how long you work on it if you make it yourself, it will never be entirely free of imperfections.

Refractor telescopes work on the principle of bending light to a point of focus. Because of this, they tend to act like a prism. The white light entering the optical system separates into its component wavelengths by the time it arrives at the focus. In a system with a simple lens, one image is formed by the red light inside the point of focus while another image is formed by the blue light just outside the focus. This is called *chromatic aberration* (Fig. 4-4) and is the bane of the refractor. Looking at anything in a system like this tends to cause headaches because everything is surrounded by halos of color. Reflectors don't have this problem because the mirror reflects all the light equally, regardless of wavelength.

All optical systems are subject to other forms of aberration. *Spherical* aberration (Fig. 4-5) is caused when the light from a lens or mirror does not reach the same

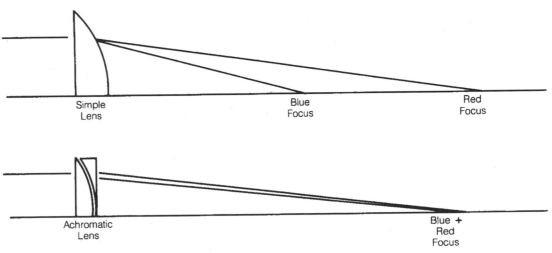

FIGURE 4-4   *Chromatic aberration brings light of different wavelengths to different points of focus. A doublet, or achromatic lens, corrects this.*

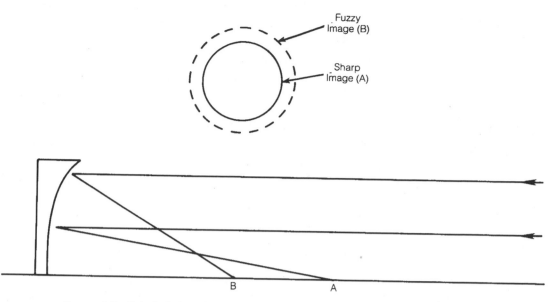

FIGURE 4-5 *Spherical aberration causes part of the image to come to a sharp focus while the rest of the image is soft and fuzzy.*

focus. An image produced by a mirror or lens with spherical aberration will appear as a series of overlapping images. *Coma* (Fig. 4-6) is an aberration that causes the images of a star to look like a tiny fan. Finally, *astigmatism* (Fig. 4-7) occurs when the

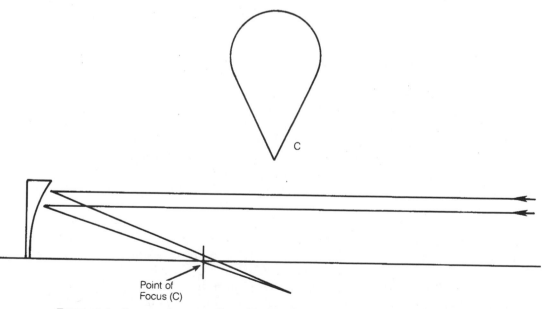

FIGURE 4-6 *Coma produces a small, conelike image because the focus is offset from the optical axis. The point of the cone always points toward the optical axis.*

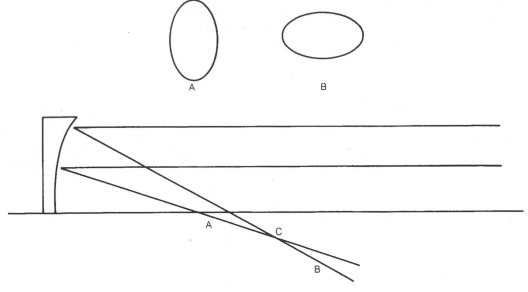

FIGURE 4-7 *Astigmatism produces oval-shaped images.*

lens or mirror is unable to bring the image to a single focus. Instead the image is formed in front of and behind the point of focus and appears blurred and indistinct.

TELESCOPE TYPES: THE REFRACTOR

When you say "telescope," an image of a refractor pops into most people's mind (Fig. 4-8). The image of the smooth, long lines of a massive refractor telescope sitting on its pier and mount in an observatory has long been the symbol of astronomy to most people.

The refractor has changed quite a bit since astronomer Galileo Galilei's time in the seventeenth century—but the operating principle has remained the same. Galileo used a simple lens system: The first lens gathered the light and focused it; then a second lens took up the duties of focus and magnification. His telescope was small, never achieving a magnification over 30 power. Today, a fairly straight-line descendent of Galileo's telescope survives in the common field glass, which uses a simple lens.

A telescope that uses a single, simple lens has one major flaw: chromatic aberration. This gives the image a halo effect—bright objects surrounded by rings of color. This is caused when the differing wavelengths of light cannot be brought to a single point of focus. In the early days of astronomy, chromatic aberration problems were overcome through the use of telescopes with extremely long focal lengths. Chromatic aberration can be minimized in a simple lens by using a focal length that moves the focal point farther away from the lens. Charles Messier—the astronomer who around 1774 catalogued a list of objects that bears his name—used telescopes with focal lengths of 60 feet. In the mid-seventeenth century, the German astronomer Johannes Hevelius built a telescope 150 feet long—so long that he could barely use it.

FIGURE 4-8 *Refractor.*

In 1757, the English physicist John Dolland discovered the basic principle of the *achromatic lens*, or *achromat*. When two lenses made of different types of glass were placed together, each element canceled the chromatic aberration of the other. Working in the early 1800s, the German physicist and astronomer Joseph von Fraunhofer discovered that chromatic aberration could be further reduced if the rear surface of the front element was more strongly curved than the front surface. He also perfected the spacing of the two elements and the achromat, or modern telescope lens, was born. Lenses produced today could have been designed by Dolland and Fraunhofer.

State-of-the-art optics today calls for the use of another lens in the design of refractor objectives. This creates a lens with three elements called an *apochromatic* lens. With the apochromatic lens, chromatic aberration is reduced to the point where it is no longer a problem in the telescope's operation.

The refractor telescopes available to today's amateur represent a vast improvement over the instruments used by some of the pioneering observers of the science. They require little if any maintenance, and their sealed tubes keep out dust and air currents that might affect the image. The alignment, or collimation, of the optical system is simple to adjust and relatively permanent.

Refractors have always been considered best for observing the Moon, the planets, and double stars. This is because the longer focal lengths found in most refractors make for a very dark field of view and allow for the use of high magnification.

These factors all contribute to the crisp, sharp images of bright objects that can be seen through the refractor.

On the other hand, it is difficult to view deep-space objects with a refractors. Deep-sky objects and objects with a low surface brightness require the use of instruments with faster focal ratios and brighter fields. A number of companies make these refractors-and a telescope with a 5- or 6-inch aperture and a short, fast focal ratio is considered the prime visual instrument for comet hunting. Leslie Pettier, the famous amateur astronomer, discovered 10 comets using a 6-inch f/8 refractor telescope.

Another major drawback of the refractor telescope is its cost. A good instrument with a well-corrected, 80-mm (3.1-in.) objective and a good, sturdy mount can cost well over $200 per inch. There are a number of companies that produce what has become a standard instrument in the refractor class. This is the 60-mm, an instrument that introduced many amateurs to the wonders of the heavens. Unfortunately, some of these manufacturers are less than strict when it comes to quality control. These instruments can be found at toy stores, department stores, and discount houses across the country for between $100 and $150. The amateur who is looking at an instrument like this would be wise to keep a wary eye on the quality of the instrument, especially its mount. A telescope will be no good to you if the mount is wobbly.

TELESCOPE TYPES: THE REFLECTOR

About 56 years after Galileo introduced the world to the telescope and wonders of the universe, another great scientific mind was considering how to improve on that work. Physicist Sir Isaac Newton knew about the chromatic aberration caused by the unequal focus of the refractor lens. He also knew that one of the properties of a mirror was the equal focus of all the light striking it, regardless of the light's wavelength. He devised an instrument that took advantage of that fact and created the instrument we know today as the *newtonian reflector* (Fig. 4-9).

The reflector telescope uses a mirror to bring light to a point of focus. The surface of the mirror has to be a specific shape-a paraboloid-for the scope to work properly. Amateur astronomers have been producing homemade mirrors for their telescopes for years. It is a fun and easy project for anyone interested in combining craftsmanship and science.

Because it is a simple instrument and easy to build, the reflector telescope is the favorite of the amateur astronomer. Newtonians are readily available from a number of manufacturers. They come in a much wider range of aperture sizes than the refractor and, owing to their ease of manufacture, are much more reasonable in cost. A newtonian with a 4.5-inch (114-mm) mirror and a solid equatorial mount can be purchased for about $300.

Reflector telescopes are free of the chromatic aberration that plagues the refractor telescope, but they have their own set of optical problems. The parabolic surface of the objective, called the primary mirror, must be made to a very fine set of tolerances. The mirror used in a reflector is subject to the same optical imperfections as a lens, with the exception of chromatic aberration. The major aberrations found in reflectors are spherical and coma. These are caused by poor optical workmanship during the figuring of the mirror. The quality of the mirrors used by commercial

FIGURE 4-9 *Newtonian reflector.*

telescope makers is good, but the constraints of the modern assembly line do not facilitate the care and time that most amateur telescope makers lavish on their instruments.

There are two major types of reflector telescopes: the newtonian and the cassegrain. In the newtonian (Fig. 4-10), light from the primary mirror is reflected off a second mirror, called the *diagonal*. Light travels from the diagonal into the eyepiece and then into the eye. Each time the light bounces off one of these surfaces a little of it is lost. Incoming light must also travel around the diagonal mirror and its supports, which causes more light loss. The amount of light that finally reaches the eye can be anywhere from 65 to 85 percent of the light that first entered the telescope. The focal length of a newtonian reflector is normally between f/5 and f/8. At the low end, an f/5 telescope makes a good deep-sky telescope if used under a dark sky. The long and more traditional f/8 gives a darker field of view and allows for a much sharper image.

Both the newtonian and cassegrain systems use a figured paraboloid for the primary mirror. The cassegrain, however, has a secondary mirror that reflects the light back toward a hole drilled in the center of the primary mirror. This is called a *folded system* because the light path is folded back onto itself. The secondary mirror in the cassegrain is not the flat type that one finds in the newtonian. It is a specially figured convex mirror that acts to increase the focal ratio of the system by three to five times. A cassegrain telescope with a primary mirror of f/5 becomes an instrument with a final focal ratio between f/15 and f/25, depending on its secondary mirror. This makes for a very compact telescope. If the primary is a 6-inch mirror, an f/20 focal

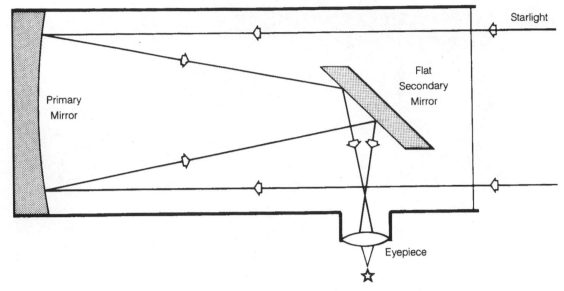

Starlight

Flat
Secondary
Mirror

Primary
Mirror

Eyepiece

FIGURE 4-10 *Light path through a newtonian reflector. (Courtesy Cosmic Connections, Inc.)*

ratio would be 120 inches long. That would be an awfully long tube for a newtonian, but the cassegrain system folds that into a tube only 30 inches long.

Another popular variation on the reflector telescope is the instrument known as the rich-field telescope (RFT). Built with focal lengths faster than f/5, these instruments give breathtaking views of deep-sky objects such as star clusters and gaseous nebulae. Because of their fast speeds these instruments require a large diagonal mirror, which causes a greater than normal light loss. The popular dobsonian design of the 1980s combines the speed of an RFT with a large mirror and smooth mounting to make for a good instrument.

TELESCOPE TYPES: THE CATADIOPTRIC

The *catadioptric* telescope, or *cat*, has been called the newest wave in telescope optics. The cat combines the mirror system of the reflector and a lens that acts as a corrector plate to eliminate imperfections in the mirror (Fig. 4-11). These instruments have been commercially available since the 1970s, but amateur telescope makers have been making catadioptric telescopes since the 1940s.

Most of the catadioptric systems available today are known as schmidt-cassegrain telescopes (SCTs). Built in sizes ranging from 4 to 22 inches, they take advantage of the small size of the cassegrain design by using a very fast primary mirror, usually f/2, and a corrector plate to eliminate the optical aberrations found in mirrors of that speed.

The greatest advantage of the SCT is its portability. An 8-inch refractor would be impossible to store in a closet. An 8-inch reflector can fit in a closet, but not much else will fit in there with it. An 8-inch SCT, however, fits easily in a case that can

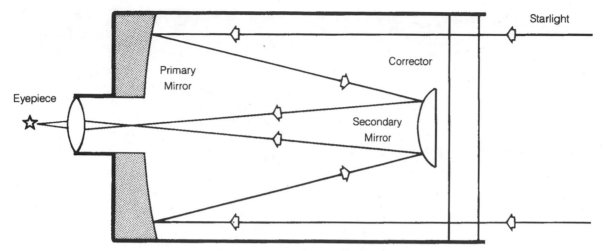

FIGURE 4-11 *Light path through a schmidt-cassegrain telescope (SCT). (Courtesy Cosmic Connections, Inc.)*

reside in a corner of the closet, leaving the rest of the space free. With the SCT it is no longer a chore to take the telescope out for an evening of observing outside the city under dark-sky conditions. The development of the SCT has allowed the amateur to take to the roadways in search of good observing conditions.

In many ways, SCTs are excellent all-around telescopes. If properly maintained they can give very good definition when used with bright extended objects such as the Moon and planets. But because it has such a large secondary mirror, the SCT suffers from more light loss than a newtonian reflector. The SCT is also subject to a loss of contrast between the image and the background sky because the mirror is perforated to admit the light from the secondary. This light enters the mirror along with a lot of stray light that randomly enters the telescope. This loss of contrast can be remedied by using a baffle at the mirror to screen the extraneous light. All in all, if there is a telescope that comes close to being a universal instrument, capable of being used in all phases of astronomy, it is the SCT. An good SCT telescope on a fork-type equatorial mount can cost about $250 per inch.

A number of companies also produce 6- and 8-inch variations of the classic newtonian reflector using a very fast mirror combined with a corrector plate in a design called the schmidt-newtonian. Another version of the cat is the maksutov. Smaller than other cats, available in 3.5- to 7-inch apertures, maksutovs use a different type of corrector plate and primary mirror that can produce images of the Moon and planets that rival those from a refractor.

OTHER FACTORS IN SELECTING A TELESCOPE

There is more to selecting a telescope than picking a color and kicking the tires (Fig. 4-12. 4-13, and 4-14). A quality telescope is a big investment, and it is important for

FIGURE 4-12 *A commercially produced 4.25-inch Rich-Field-Telescope (RFT).*

the amateur to consider all factors before making a selection. In addition to concerns about aperture size, resolving power, and the strong and weak points of individual telescope types, a few general questions must be answered. The three S's-subject, site, and sky-all combine to help narrow the choices available to the amateur astronomer.

The choice of subject is the most important consideration. Just what do you want to look at with the telescope? Objects like the Moon, planets, and double stars require high power, sharp resolution, and a dark field. A telescope suitable for studying these objects, however, will fall short when turned on deep-sky objects. On the

FIGURE 4-13 *A Dobsonian telescope*

other hand, an instrument that gathers enough light to study the fine structure of a diffuse nebula will be woefully inadequate for viewing the Moon.

Unfortunately there is no such thing as the perfect overall telescope. Regardless of your final choice, you are going to have to give up something along the way.

The site from which you will be observing will also have something to do with the type of telescope you select. An observer living in the middle of a major city, surrounded by light pollution, is going to have a different set of observing priorities than a person living 150 miles outside that city. For the observer at the bottom of a pool of light pollution, the best objects to view may be those with a high surface brightness such as the Moon, the planets, and brighter stars.

It is also important to consider the transportability factor of the telescope. If you will have to move the instrument to the observing site, it is best to get a light, compact telescope. If you are going to observe from your backyard and can use the garage for storage, size may not mean that much to you.

Finally, if you are going to buy a telescope, get to know the dealer at your local astronomy shop. There is usually at least one in each large city. These people usually active amateurs themselves and can guide you along the road of equipment acquisition by sharing their experience. If you plan to build a telescope yourself, look both to your local dealer and the members of your local astronomy club for help. They are only too

FIGURE 4-14 *A reflector telescope.*

pleased to see a new member join the fold and can be a great help to a beginning amateur astronomer. And if you live in a small town, find the nearest astronomy club—again, they are only too pleased to help you get acquainted with the universe.

TAKING YOUR SCOPE FOR A TEST DRIVE

The best way to check the performance of any telescope is to use it. It would be nice if you could take the telescope for a test drive before you buy it, but this is impractical. Once you have the telescope home and assembled, take a few minutes to get acquainted with it. Get to know how the instrument will feel while it is light outside. See how far you can reach while at the eyepiece. Are the slow-motion controls readily accessible? If the telescope has a star diagonal, can you use a stool or chair to get comfortable? If you can, how far toward the horizon is this useful? Practice moving the telescope through short arcs to get the feel of its ease of motion. Your next step should be to get it outside for a few simple performance tests.

ALIGNING THE OPTICS

One of the biggest bugaboos confronting the beginner is the need to align the optics of the telescope and then to keep them in alignment. After you have the telescope set up and are acquainted with it, it is time to *collimate* the optical system. Don't let that big word intimidate you. All you are going to do is make sure the light entering the telescope reaches your eyes properly.

If you have a refractor telescope it is easy. Because the refractor has only one optical element, the objective lens, it is the easiest to align. Most refractors sold today come with the objective in perfect collimation, and, unless you are planning to use the telescope as a baseball bat, it will probably stay in alignment indefinitely.

A reflector telescope is another matter. Here you have two elements that must line up perfectly to pass the image along to the eye. The primary mirror is mounted in a cell and can be jarred out of alignment with a solid thump. The diagonal is suspended at the end of a single stalk of thin metal or from the vanes of a spider mount. Either mounting system is susceptible to movement or vibrations that can knock it out of alignment. The actual procedure for collimating a reflector is not that difficult, and most of the work can be done in full daylight.

The first step is to get a small pinhole sight and a set of small screwdrivers and wrenches. There are three ways you can get the necessary pinhole sight. You can buy one for about $30. The main drawback to this method is the $30. You can take the lens system out of an old, high-power eyepiece you no longer use. But if you are just starting out it is doubtful you have any old eyepieces laying around, so let's move on to the third option.

The easiest way to secure this precious device is to take a few pictures of your new telescope. The canister the 35-mm film came in will suit our purposes nicely, and you will have a memento to show your friends. With the canister firmly in hand, remove the lid and look at its base. There is a small dimple in the center where the canister was attached to a spur when it was manufactured. Drill a small hole, centering on the dimple. The hole should not be larger than 1/16 inch. Now take the canister and turn it over. See the lip where the lid fit? Carefully cut that off with a sharp knife or razor blade. The resulting lipless, hole-in-the-bottom film canister will now fit into the telescope's focuser. Our newly constructed "collimation eyepiece" is now ready for use.

With the telescope pointed at the sky, rack the focuser in as far as it will go and look into the pinhole sight. You will see, from the outside in, the edge of the focuser, the diagonal's holder, the outline of the diagonal itself, the reflection of the primary mirror, and the reflection of the diagonal reflected in the primary (Fig. 4-15). The first thing you want to do is center the diagonal in the telescope tube. If the diagonal is mounted on a stalk, the stalk will be connected to the focusing mount by means of a set screw. Loosen this screw and sight down the tube to position the diagonal holder in the center of the tube. A spider mount will have the holder fastened to the

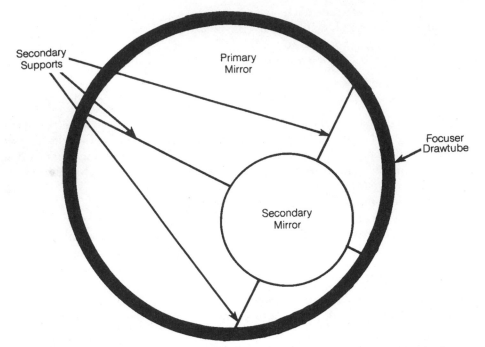

Secondary
Supports

Primary
Mirror

Focuser
Drawtube

Secondary
Mirror

FIGURE 4-15 *Begin to collimate your optics by peeking through the drawtube with no eyepiece in place.*

mount vanes by means of a bolt. You can loosen the bolt and adjust the diagonal holder in the same way as in the stalk method.

Now go back to the pinhole for our next step. You want to position the reflection of the primary mirror in the center of the diagonal (Fig. 4-16). You may have to bend the stalk slightly to do this but don't worry—it is built to take this treatment. With a spider mount, you need to adjust a number of small setscrews to move the diagonal mirror. Make your adjustments until the primary mirror, complete with the visible section of the mirror cell, is centered in the diagonal mirror.

Now it's time to deal with the primary mirror. At the back of the mirror cell you will find a number of knurled bolts, usually three. Go back to the pinhole sight and check the position of the reflection of the diagonal. What you want to do is move this reflection until it is centered in the reflection of the primary mirror (Fig. 4-17). Check the image, then go to the primary and turn the appropriate screw. Repeat this until the image is centered.

The next thing to do is align the finderscope. If you aligned it before you started moving the primary, you can be sure that the finder needs readjustment. You can do this during the day so you don't have to fumble with any adjustments in the dark. Select an object at least a quarter of a mile away, the farther away the better, and sight along the telescope tube until it is in the field of view of the main telescope. Put in your highest-power eyepiece and center the image in the field. Now look through the finder and see how far off the target is from the center of the crosshairs. Adjust the image until the crosshairs are centered on the target using the setscrews around

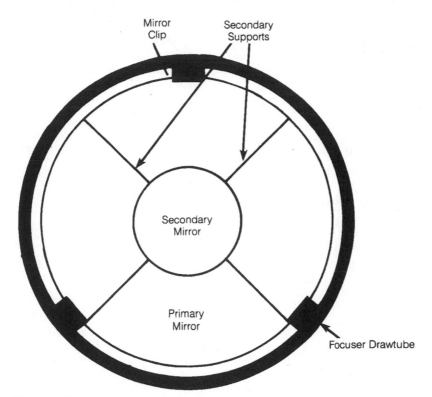

FIGURE 4-16 *This is what you will see when the mirror is properly centered.*

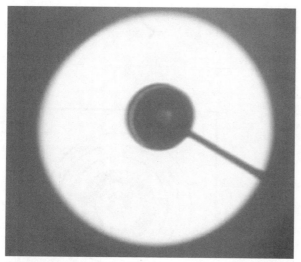

FIGURE 4-17 *A properly collimated system.*

the finder's mounting ring. Now select another object that is farther away and aim the telescope using the finder. If the finder is properly aligned, the target should be centered under the crosshairs and in the main field of view. If it is off center, readjust the setscrews until the target is centered. There is nothing more annoying than trying to find an object with a finder that is not properly aligned.

When the Sun has set, find a bright star and bring it into view. Now it is time to fine-tune the system. Using a medium-power eyepiece, throw the star slightly out of focus. You will see the star's diffraction rings. If you have properly aligned the optics, the rings will be concentric (Fig. 4-18). Most times, however, the rings will be slightly off and you will have to make a few more subtle turns on the primary cell setscrews (Fig. 4-19) to bring the rings into line.

SCT telescopes are easier to align. Since there is no way for you to get at the primary mirror, all of your adjustments will be made with the secondary mirror. There are three small screws around the edge of the secondary mount. These are all you will be using. If there is a large bolt in the center of the secondary mirror—leave it alone. Remove the star diagonal and put a medium-power eyepiece in the focusing tube. Find a bright star and rack the focus out until a dark disk appears at the center of the out-of-focus image. This is the silhouette of the secondary mirror caused by the incoming light from the star. The dark spot should be centered in a ring of light. If it isn't, adjust the screw opposite the direction that the dark spot is off-center. For example, if the spot is left of center turn the screw to the right. After this rough

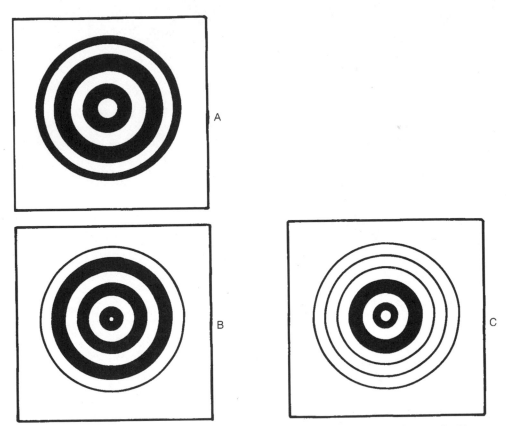

FIGURE 4-18 *Good star images are the product of proper collimation. Star image A is what you should see with the image in focus. Images B and C are outside and inside of focus, respectively.*

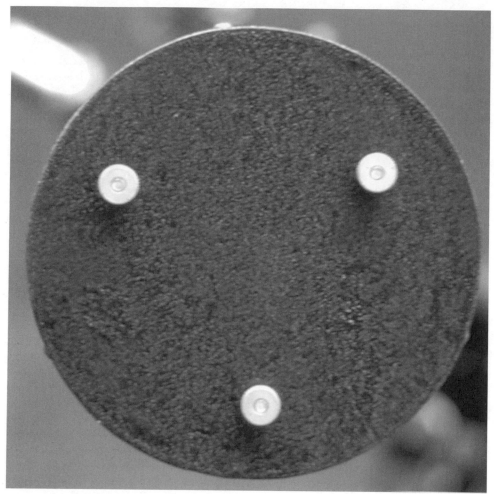

FIGURE 4-19 *The setscrews in a newtonian telescope.*

alignment is made, put in a high-power eyepiece and find a third-magnitude star. Center it in the telescope's field of view and slowly rack the image out of focus. The diffraction rings should be concentric. If not, make any necessary adjustments to the secondary mount.

STAR TESTING THE TELESCOPE

It's not a bad idea to give the telescope a quick star test whenever you start an observing session. It's easy to do and it can alert you to any number of problems.

Set the telescope up in the daylight and make sure the optics are properly aligned. Check the field of view. Put in a low-power eyepiece and find an object that fills the field. The object should be in focus across the entire field. If the center of the target is crisp but the focus becomes fuzzy at the edges, the field is not flat. Switch to another eyepiece. If the problem disappears, the first eyepiece has a flaw.

If it persists, the optical system may still be out of alignment. The image should also be uniform. Any bulges or pinches in the image are a sign of distortion and indicate a problem.

Take a few minutes to get acquainted with the telescope in the dark. Select a second-magnitude star, such as Polaris. It is a good test object because it doesn't move. Using a high-power eyepiece bring the image to sharp focus and then slowly move the focus of the instrument back and forth. When it is in focus the star will form a small disk-the airy disk described in Chapter 4.

As you move the focus inside or outside the focal plane the image should expand a little but look the same at either side of the focal point. In a reflector the image formed inside the focal point should be surrounded by a series of rings—the diffraction rings. Outside the focus a dark spot should form in the center of the image. This is caused by the shadow of the secondary mirror.

In a refractor the image will show the effects of chromatic aberration. At the point of best focus, the image will have a yellow tint with a purple halo. Inside this focus the disk will have a reddish-purple color at the edge. As you move outside the focus, the central image will become a red disk. As you move further out from the focus, a green-yellow edge will appear until a bluish flare forms over the central image.

In all types of telescopes the brightness of the image should be the same across the disk. If the image is brighter at the edges than the center, or vice versa, the objective is showing spherical aberration. If the diffraction rings flip as you move through the focal point the system has astigmatism. If the image is fan-shaped, a coma is present.

A star test can show you if your optics are poorly made or designed. If you took the time to make the telescope you are testing and any of these aberrations shows up, it is time to return to the optical shop and perform a bit of optical surgery on your patient. If these things show up in a commercially produced instrument and you cannot get rid of them no matter how much you collimate the optics—or yell and scream—take the instrument back.

5

EYEPIECES AND MOUNTINGS

A telescope with good optics and a lousy eyepiece is a lousy telescope. . . .

*A telescope with good optics, a good eyepiece, and a lousy mount
is even worse!*

OVERHEARD AT ASTROFEST 1988

Many amateurs spend large amounts of money making sure that the telescope they use has the finest optics and the sturdiest mount available. Then they go out and buy bargain basement eyepieces and wonder why the images are fuzzy and out of focus. The telescopes used by both professional and amateur astronomers are made up of two interrelated optical systems. The primary optical train consists of the objective and any secondary mirrors involved in passing the image to the eye. The other optical system is the eyepiece, something most beginners overlook.

EYEPIECES

The eyepiece is a collection of lenses that magnify the image formed by the telescope's objective. It is made up of two main lenses: the field lens, which faces the telescope objective, and the eye lens, which lies nearest the eye.

Eyepieces come in a variety of types, shapes, and sizes (Fig. 5-1). An eyepiece is usually identified by its focal length and a letter designating its type. For example, an 18-mm Ke is a kellner lens with an 18-mm focal length; a 6-mm Or is an orthoscopic lens with a 6-mm focal length.

The focal length of the eyepiece determines two important factors. First and most important, it determines the size of the exit pupil of the eyepiece. Dividing the eyepiece's focal length by the focal ratio of the telescope gives you the size of the exit

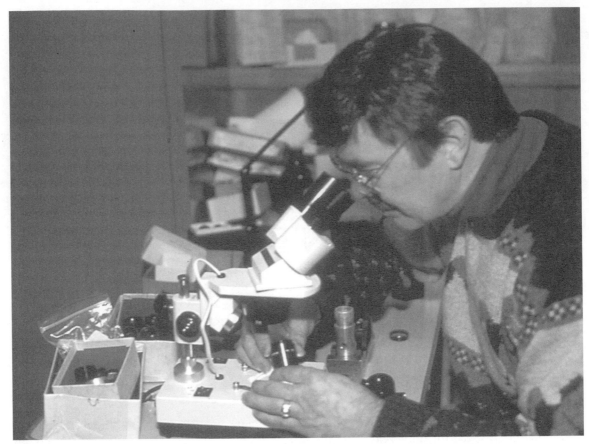

FIGURE 5-1 *Telescope eyepieces.*

pupil for that particular eyepiece. For example, if you want to know the size of the exit pupil of a 10-mm eyepiece when it is used with a Meade 4-inch schmidt-cassegrain telescope (SCT), divide the focal length of the eyepiece by the focal ratio of the telescope and you will get an exit pupil of 1 mm (10 mm/10). The lens will have the same size exit pupil when used with any telescope with a focal ratio of f/10. If you put that same lens in a telescope with a focal ratio of f/5, the exit pupil becomes 2 mm (10 mm/5).

Secondly, the focal length of the eyepiece tells you the amount of magnification the eyepiece imparts on the image passing through it. To obtain this figure, the focal length, not the ratio, of the telescope is divided by the focal length of the lens with both distances expressed in the same terms, either millimeters (mm) or inches (in.). The 10-mm eyepiece in the Meade SCT magnifies the image 100 times (1000 mm/10 mm), written as 100× or ×100.

Unfortunately, when most people think of a telescope, they think of its magnifying power. As often as not, the first words out of a person's mouth when a telescope comes into view is something like "what power is that thing?" It doesn't do any good to magnify an image to the point where it becomes unclear and blurred. This wastes the real power of the telescope-its ability to see small details.

The eyepiece should make full use of the resolving power of the telescope. Remember that when the magnification is increased the field of view is decreased. Increased magnification not only increases the apparent size of the image, it also increases the apparent speed of the Earth's rotation, causing the image to move faster across the field of view. Increased magnification also magnifies each little tremor that passes through the optical train. For instance, a puff of wind can make the image shake and jump until it becomes impossible to see anything of interest.

EYEPIECE ABERRATIONS

Because light must travel through the lenses that make up an eyepiece, the eyepiece is subject to the same aberrations as a telescope's objective. An eyepiece that causes the image of a star to flare as you shift your center of vision suffers from spherical aberration. Caused by the uneven focus of light, this aberration makes it difficult to focus. By their very nature lenses are subject to chromatic aberration, that annoying habit of adding a rainbow halo of color to a bright object. Distortion in a lens causes a straight line to curve as it approaches the edges of the field. *Field curvature* creates a lack of sharpness throughout the field of view, with the center being in sharp focus while the edges are fuzzy and indistinct. Astigmatism in an eyepiece will cause the image of a star to appear as a cross or square the further away it is from the center of the field. Finally, some eyepieces are made up of a number of optical surfaces. The image may reflect off these surfaces as it passes through the eyepiece, causing ghost images to form. Tracking the source of an aberration can be difficult because both the eyepiece and the objective may suffer from the same set of defects.

EYEPIECE BARRELS

The eyepieces used on a telescope are determined by the barrel size a telescope can accept. The barrel of the eyepiece comes in three sizes. The 24.5-mm (or 0.965-inch) has been called the Japan standard size. Many of the small telescopes found in department and discount stores come with eyepieces of this size, although the availability of useable eyepieces is limited.

For most amateurs, the most popular barrel size, known as the American standard, is the 1.25-inch-making it easy to find eyepieces in this size. The popularity of large-mirror, short-focal ratio dobsonians has also lead to a proliferation of 2-inch-barrel eyepieces. These eyepieces are becoming more affordable, and the variety and selection continues to grow. Certain telescope dealers sell focusers that have adjustable 1.25- and 2-inch sleeves to hold both eyepieces-and there are even interchangeable dual 1.25- and 2-inch barrels.

An adapter will let you use a smaller-diameter barrel with a larger-diameter focuser. You can use a 1.25-inch lens in both a 0.965-inch and a 2-inch focuser, but in most cases you cannot use a 2-inch barrel lens in a 1.25-inch focuser unless it is of special design. You must be careful, however, when trying an adapter between the 1.25-inch and 0.965-inch barrels-or any of the interchangeable adapters. The focus of the 1.25-inch lens may be slightly longer than the travel of the focusing tube on some small refractors. Try to find out the amount of extra travel that the larger lens will require. A number of companies make and distribute star diagonals that are also adapters.

TYPES OF EYEPIECES

Eyepieces are not all the same. In fact, in the past decade, the number and variety of eyepieces has dramatically increased—and we will only mention some of the lenses in this section. Many companies in the United States and Japan (among other countries) are designing larger, more complex eyepieces with superior coatings that enhance images—all within a reasonable price range of $75 to $250; superior lenses, of course, cost several hundred dollars. Most beginning amateur astronomers start with a good, less expensive lens. Then, after some experience-and checking out what other amateur astronomers use at star parties and astronomy clubs or contacting amateur groups on the Internet-look into more expensive lenses.

We will start with the *huygens* (H) eyepiece, the type supplied with many refractor telescopes sold in department stores. A cheap and easy eyepiece to produce, the huygens is best suited for the slow, longer focal ratios like f/15. While it produces an adequate field of about 30 degrees without ghost images, the huygens is plagued by other optical problems. The worst of these is its tendency toward severe spherical aberration when used with faster focal ratio systems. The huygens eyepieces are also subject to distortion, curvature of field, and chromatic aberration. These aberrations can be minimized, however, by using a long-focal-length system.

These eyepieces are difficult to find because they no longer come with the popular instruments produced in this country. Even the imported small scopes are moving away from this eyepiece type, but they can still be found if a person looks hard enough. If you have a telescope that comes with a huygens, don't throw the eyepiece away. They do have their uses-as you will see. Instead, wrap it carefully and put it in the back of the drawer.

Another lens that comes with many small telescopes is the *ramsden* (R). Just as easy and cheap to produce as the huygens, the ramsden uses the same types of lenses as the huygens. The improvement made to the ramsden is that these lenses now face each other, thus reducing some of the defects found in the huygens. Spherical aberration is greatly reduced. Chromatic aberration, however, especially at the edge of the field is still apparent.

This eyepiece also has its own set of problems. The eye relief is short, requiring the observer to put his or her eye very close to the eye lens. The field of view is smaller than that of the huygens, and ghost images are more noticeable. Ramsdens can be used with slightly faster optical systems, down to f/8, but they should not be used with an instrument like a rich-field telescope (RFT). The ramsden seems to be the eyepiece of choice for the importers. Most companies offer their scopes with a couple of ramsden eyepieces, usually a 12-mm and a 4-mm. Like the huygens, the ramsden will have its uses-so don't discard it.

The *kellner* (K or Ke) is an improvement on both the huygens and the ramsden. Often called an achromatic ramsden, the kellner uses an achromatic doublet as an eye lens. Like the achromatic objective of a refractor, the qualities of the two types of glass that make up the doublet cancel out the aberrations each causes. The immediate advantage to this system is the reduction of chromatic aberration. While astigmatism and curvature of field are still apparent, they are greatly reduced. The large field of view, close to 50 degrees, and good eye relief make the kellner an excellent eyepiece for low-power viewing. These eyepieces are subject to ghost images, however, and to unfocused images near the edge of the field. Kellners are made by a number of companies and come in all barrel sizes. They come in focal lengths from 6 to 40 mm and start from about $50, though they are often discounted.

In the 1970s, Edmund Scientific Co. commissioned Dr. David Rank to develop an improved version of the classic kellner eyepiece. Called the *rank-kellner eyepiece* (RKE), it consists of an achromatic field lens and a single double-convex eye lens. The production is overseen by the company to exacting standards. The result is a lens with a very sharp field of 45 degrees. These lenses give the amateur on a budget a chance to obtain excellent eyepieces at a reasonable cost. The RKE eyepieces list at about $50 direct from Edmund Scientific and range in size from 8 to 28 mm in the American standard barrel size.

The *orthoscopic* eyepiece, or ortho, was once considered the best all-around eyepiece—at least until its narrower field lost out to some newer designs. It has a three-element field lens and a plano-convex eye lens that produces a field of 45 degrees with good eye relief, excellent sharpness and contrast, and little chromatic aberration. The eyepiece is good for viewing the planets and the Moon. However, it suffers from astigmatism and a strongly curved field.

Orthos are made by everyone. Many manufacturers contract out the construction of this type of lens to small Japanese firms. The result is that every company has a line of orthos carrying their name. Orthoscopic eyepieces come in all barrel sizes and range in focal length from 4 to 25 mm. Prices vary over a wide range starting at $50 for a private label 0.965-inch barrel lens and going all the way to $250 for a 2-inch barrel, although these lenses are frequently discounted. The 1.25-inch lens is priced between $50 and $80. A diligent check of the price lists in the major magazines will reveal a wide range of discounted prices.

The *plossl* eyepiece design uses two identical achromats to give good eye relief and a wide field of view. It is less troubled by astigmatism than the orthoscopic eyepiece, but it has a slightly higher level of distortion. Plossls can be found in the American standard barrel size in focal lengths from 3.8 to 45 mm, and from 40 to 56 mm in 2-inch barrels. Construction of these eyepieces has been turned over to foreign jobbers. As a result, many private-label plossls exist. Prices for the 1.25-inch barrels range from about $80 to $120, depending on the size, and can reach up to $250 for the big 2-inch sizes.

The *erfle* is often the eyepiece preferred by the observer who wants a wide field with good definition. The lens placement varies among manufacturers, but usually includes achromatic field and eye lenses, which are separated by a third lens, most often a double-convex. A popular lens for deep-sky observing, the erfle can be found in the 1.25-inch barrel with a size range from 20 to 40 mm. Prices for a good erfle are right around $90.

The *konig* eyepiece is a variation of the erfle. It is a bit simpler in design but produces an excellent image and field definition.

Lanthanum eyepieces (yes, the same name as the rare earth chemical element) are produced by several companies. Most Lanthanums have a narrow field of view, about 45 to 50 degrees depending on the eyepiece size, and they have a much longer eye relief than most eyepieces—a 20-mm eye relief. They also show a trace of false color at the edges, but they are low in price, ranging from $130 to $150. The lenses range in size from 2.5 to 25 mm. There are also Lanthanum superwide eyepieces, with apparent fields of about 65 degrees—still with the 20-mm eye relief. The prices range from about $205 to $225; many fit 1.25- and 2-inch barrels.

The wide-field (ultrawide or wide angle) eyepieces compete with erfles in definition, field size, and eye relief. Most of them offer several lens elements within the eyepiece, with fields of view that can reach over 80 degrees. The only drawback, although it is hardly noticeable for most observers, is the light loss from such a wide field of view. In most cases, the eyepiece quality is high—and so is the price.

In the past decade or so, wider-field eyepiece offerings have continued to grow. One good example is a lifelong amateur astronomer turned professional optical designer, Al Nagler. In the 1980s, Nagler's company, TeleVue, brought out a line of *nagler* eyepieces. The naglers have become a mainstay for deep-sky observers because they offer wide fields and excellent definition. In fact, the eyepieces are known for their wide field of view: Most eyepiece designs concentrate on the human eye's 50-degree field of view—thus many eyepieces, such as the kellners and RKEs, are limited to 40- to 45-degree fields of view. The naglers offer an amazing 82-degree view, with the entire field—practically to the edges—in sharp view (and observers without a clockdrive don't have to move the scope as much because of the wider view). Although we as humans can't take in such a wide field, it's easy just to move your eye or head around to take in the view. The eyepiece also maintains a much larger lens to look through and a much longer eye relief than you would expect from the given focal lengths.

There are a few drawbacks, including the size and weight, which may mean that you have to rebalance your telescope when you put in different lenses. Another drawback, but only for the 9-, 11-, and 13-mm, is a kidney bean-shaped distortion as you move your eye around; it's worse in daylight observing, and TeleVue mentions that the 13-mm is "not suitable" for daytime viewing. It bothers some people, but not others. The Nagler type 2 series "corrects" for this, but the eye relief is shorter in these eyepieces. The sizes range from 4.8 to 20 mm (the 17-mm and 20-mm were replaced by the Nagler type 4 eyepieces of the same size, but you may still find some of the discontinued lenses). The prices are a bit hefty: from about $205 list price (about $170–190 street price) for a 4.8-mm to $535 list ($425–450 street price) for a 20-mm. But many amateurs say that they use just one nagler in the place of many other eyepieces.

The number and types of naglers continues to grow. The Nagler Type 4 was introduced around 1999, with additional features such as about 17 mm of eye relief and less "pincushion" distortion (this makes straight objects look bent at the field's edge—not necessarily bad for astronomical viewing, but something to be aware of), and some of the lenses are not as heavy.

The *Panoptic* eyepieces are also enjoyed by many observers and are part of the TeleVue line. These eyepieces have great eye relief and about a 68-degree field of view. They range in size from 15 mm to 35 mm, with somewhat high prices: about $210 for the 15-mm to $365 for the 35-mm (2-inch barrel). Many of these eyepieces also have pincushion distortion, which may not be bad for astronomical viewing, but during the day makes straight objects such as trees look bent at the field of view's edge.

The *radian* eyepieces (in mathematics, a radian equals 57.3 degrees—the approximate field of view in a radian eyepiece) feature 20 mm of eye relief no matter what eyepiece you buy. The wider field of view is both sharp and color-free, even to the

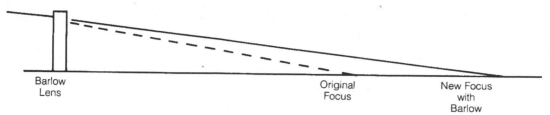

Barlow
Lens

Original
Focus

New Focus
with
Barlow

FIGURE 5-2   *The barlow lens acts to lengthen the focal length of your objective, allowing you to get a higher magnification from your eyepieces.*

edges, with a large eye lens; many observers compare the radians to the wide-field orthoscopics. Similar to many of the wider field eyepieces, the radians weigh more than the average eyepiece. They range from 14 mm down to a super-short 3 mm, with prices around $225 to $300 each.

The *barlow lens* is not really an eyepiece at all, but acts to increase the focal length of the eyepiece used (Fig. 5-2). The barlow lens increases the power of the eyepiece by two to three times. Most eyepieces are restricted to a fixed amount of magnification, but some can be changed by varying the distance between the field lens and the barlow lens. Using a barlow lens with an 18-mm eyepiece gives you the equivalent of a 9-mm with a 2× barlow or a 6-mm with a 3× barlow. The barlow effectively multiplies the number of lenses in your collection by two. If you have a 12-mm lens and a 2× barlow, you don't have to buy a 6-mm lens because you already have it when you combine the 12-mm with the barlow.

Another advantage of the barlow is that it retains or increases the eye relief of the original lens, a boon to those who wear glasses. Barlow lenses are another item that seems to be made by everyone, for every barrel size. One important thing to remember about the barlow is that the field lens should be adequate in size or the image brightness will suffer drastically. Prices for a good 2× barlow range from $120 to $180; the 3× barlows are about $120. You can even get a 2.5× RKE barlow for about $100 through Edmund Scientific.

Although most amateurs believe these are a bit frivolous, what about zoom eyepieces advertised by many eyepiece dealers, mostly for observers tired of changing lenses all the time? These eyepieces offer a range of sizes; for example, a 7-mm with an apparent 65-degree field of view zooms to a 21-mm with a 35-degree field of view. But like zoom lenses for a camera, zoom eyepieces lose some of the detail and definition. The prices range from $150 to $200. Of course, there is always the multiple eyepiece magazine, with four holders that rotate the eyepieces into viewing position, which sells for about $190.

SELECTING AN EYEPIECE

The best eyepiece for any particular observational need meets two simple qualifications: It fits your observing needs, and it fits your budget. If your interests are planetary and lunar observations, you will want an eyepiece that gives very sharp, clean images. Orthoscopics, plossls, and RKEs fit this bill of fare very nicely. If you want sweeping views of deep-sky objects such as star clusters, a lens with a wide, flat field of view like an erfle or a konig is more appropriate.

TABLE 5-1.   EXIT PUPILS

EXIT PUPIL SIZE (MM)	OBJECT TYPE VIEWED
5-7	Galaxies
	Nebulae
3-4	General viewing
	Clusters
	Planetary nebulae
2	Moon and planets
	Double stars
1	Maximum for Moon and planets
	Close doubles
0.5	Closest doubles on best nights

Source: Jeffrey L. Hunt, Cosmic Connections, Inc.

The size of the eyepiece is another thing to consider. Remember that the important feature of the eyepiece is the size of its exit pupil, not its magnification. Use the information in Table 5-1 as a guide for selecting a range of eyepieces.

If, for a simple example, you use an 80-mm refractor telescope with a focal length of 910-mm, what types of eyepieces should you have in your observing kit? The telescope's focal ratio is slightly over f/11, f/11.4 to be exact. To keep the math simple let's stick to f/11. According to Table 5-1 an eyepiece for general viewing should have an exit pupil of 3 or 4 mm. To get this exit pupil, you will need an eyepiece with a focal length of 33 mm. A 33-mm eyepiece is a awful lot of glass and is going to cost a pretty penny. A 32-mm plossl will fit your needs, but it can be expensive. Konigs are popular and less expensive. An erfle would be ideal but, for reasons unknown to me, not many places carry them. Let's say you decide to go with a 32-mm erfle and after a bit of hunting you track one down for about $60.

For viewing the Moon, planets, and double stars, the table recommends an exit pupil of 2 mm. With our telescope that would work out to a 22-mm-focal-length eyepiece. You have a variety of eyepieces that will give this exit pupil. A 22-mm plossl or a 21-mm RKE will be adequate for your needs. You decide to go for the quality of the plossl and spend $120. Total so far: $180.

When seeing is good and you want a close look at the belts on Jupiter, you will need an exit pupil of 1 mm or an 11-mm-focal-length eyepiece. You have a 12-mm orthoscopic, RKE, or private label plossl from which to choose. This time you decide to choose the 12-mm RKE for $50. Total spent on eyepieces: $230.

For those nights when the atmosphere is calm and the seeing perfect, a 0.5-mm exit pupil could be used to split doubles right at the limit of the telescope. A 5- or 6-mm-focal-length lens is what you are looking for and, again, there are many choices. You decide that for an eyepiece of this focal length an ortho is ideal. You spend another $80 for a good 6-mm lens. This brings your final total to $310.

You now have four eyepieces ranging in size from 6 to 32 mm, with exit pupils from 0.5 to 3 mm. What magnification range do you have? The 32-mm erfle gives you a magnification of ×28, the 22-mm plossl gives ×42, the 12-mm RKE ×76, and the 6-mm gives you a magnification of ×152. The magnification per inch of aperture ranges from ×9.5 per inch for the 32-mm erfle to ×50 per inch for the 6-mm ortho. These figures fall perfectly within the optimum for the telescope's best performance.

TELESCOPE MOUNTS

Exit pupil size and magnification mean nothing if the object you are looking at is constantly jumping around the field of view. Poor mountings have contributed to more than one amateur astronomer trading in the telescope for a seat in front of a wide-screen television.

A good, solid mount is an essential piece of equipment for the amateur (Figs. 5-3 and 5-4). The mount should meet these two basic requirements: It should provide solid support for the weight of the telescope; and it should be able to move easily along both the horizontal and vertical axes.

Support is the main job of the telescope mount. It must provide a stable base that will dampen any vibrations traveling through the system. A mount won't eliminate

FIGURE 5-3   *A Maksutov telescope with an altazimuth telescope mount.*

vibrations, but it should minimize them to a point where they quickly disappear. One of the primary contributors to a poor image through a small telescope is the shaky mount. The optics of a small telescope may be very good, but if the mount is flimsy the optics are almost useless.

ALTAZIMUTH MOUNTS

The type of mount responsible for all this trouble is a cheap variation of the altazimuth mount. This mount usually consists of a pillar and claw, with the telescope balanced between two points on the claw. This type of mount is so unsteady that many come with a bracing bar—which only further complicates things. When a telescope is mounted on a tall, unbraced, and rickety tripod, it is a wonder anything is seen at all.

Altazimuth mounts suffer from another shortcoming—they make it difficult to follow the course of the observed star. Because of the way stars move across the sky, a

FIGURE 5-4 *An equatorial telescope mount. (Photo taken courtesy of Don Yeier, Yeier Optics)*

telescope needs to track along two axes. Many refractor telescopes come with "slow-motion" cables that make tracking easier, though tracking with an altazimuth mount is still difficult.

This is not to say that all altazimuth mounts are unstable—they aren't. Mike did all of the observational work for this book with an 80-mm refractor telescope on an altazimuth mount. The mount is stable and offers excellent support. The concept behind the altazimuth mount is simple: provide a simple, easily constructed base for the telescope. In the nineteenth century many telescopes were carried on altazimuth mounts and many pioneering discoveries made by amateurs were made with altazimuth-mounted instruments.

A popular version of the altazimuth is the dobsonian mount used on many large reflectors. The dobsonian is designed to provide a stable platform that is still easy to move when tracking celestial objects. The mount carries the weight of the main portion of the telescope—the primary mirror and its supporting cell—inboard or between its supporting columns. This makes the mount very stable and solid.

Another advantage of the dobsonian mount is its ability to move along the two axes simultaneously with very little effort. You can push the telescope easily, owing to the use of Teflon for bearing material, and when you stop, the mount stops.

Equatorial Mounts

The best mounting for any telescope, no matter how small, is the equatorial mount. Invented by the nineteenth-century German scientist Josef von Fraunhofer, the equatorial mount is designed to follow the movement of a celestial object with a minimum of effort. The modern equatorial has a variety of shapes, each one with its own distinct advantages and disadvantages. The most popular of these is a direct descendent of Fraunhofer's mount and is called the German equatorial mount or GEM.

The altazimuth mount tracks a star by moving through two perpendicular axes. This means that for every degree of movement made by the star, the telescope and mount must make two movements. The equatorial mount simplifies this by making one of its axles parallel to an axis. This is the polar axis of the mount. It extends from the mount toward the north celestial pole. With this system one of the two axes is stationary, so the mount has to swing the telescope through only one axis—the right ascension axis. This setup also allows for the addition of a motor or clock drive, to make the movement through this axis automatic. The equatorial mount also allows for the use of setting circles on each axis to index star and object positions.

Another popular mount is an equatorial variation called the fork mount. Found on the schmidt-cassegrain telescopes (SCTs), the fork shrinks the equatorial mount to a size that will fit in a small box. The fork mount points the body of the mount toward the pole until the "tines" of the fork are parallel to the polar axis. The telescope is then cradled inboard between the tines for a vary stable configuration.

Aligning an Equatorial

One of the few advantages of the altazimuth mount is that you don't need to worry about where or how to set it up. Just step outside, plant the tripod, and you are ready to go. With an equatorial mount it isn't that easy. Unless you have the mount set on a permanent pier, you will have to align the mount each time you use the telescope.

If you don't have your mount fitted with a clock drive, you may just need a rough alignment. To do this, simply sight along the mount in the direction of Polaris. Depending on your skill, the mount should be within 5 to 10 degrees of the pole. If you only plan a short viewing session, this is adequate.

For a better alignment, first use the rough alignment technique and then clamp the telescope at 90 degrees declination. Move the entire mount until Polaris is nearly centered in the telescope's finder. This should improve the accuracy to within 2 to 3 degrees of the pole.

Many of the commercially available mounts come with a handy feature called a *polar alignment reticle*. This little gadget is designed to show you exactly where the pole is in relation to Polaris. When it is mounted in the finder of a SCT, you simply give the telescope a rough alignment by eye, line Polaris up in the finder, adjust the finder so that the pole is at the proper mark on the reticle, and you are ready to go. Many companies also offer this feature with a GEM mount. All you do is sight through the polar axis because the mount has a finder scope built in, complete with a polar alignment reticle.

FINDING YOUR WAY AROUND THE SKY

I can see it with my naked eye, but I can't find it with the telescope!

The next time you find yourself away from the lights of a city, take a minute and give the night sky a quick glance. Make it quick because looking at the night sky undimmed by light pollution is like gazing at the face of Medusa—it may be entrancing but it could turn you into stone. The seemingly countless number of stars has dazzled many budding amateurs, sending them scurrying for shelter. "How can anyone find anything in a star-choked sky?" they ask. Tracking down a star cluster in the apparent maze of stars is not that difficult. All it takes is a little inside information and a bit of practice.

COORDINATES ON THE SKY

One method of finding your way around the sky is to assign coordinates to celestial objects. This way you will know where any object is at any given time. Once you decide to do this, you will need to determine the type of coordinate system you will use.

RIGHT ASCENSION AND DECLINATION

In the fifteenth century, an enterprising Italian inventor and mapmaker named Toscanelli dal Pozzo was intrigued by the inability of navigators to find their way around the seas after they lost sight of land. He reasoned that if a grid of equally spaced lines were placed over a map, it would be easier for a ship to know where it was and where it was going. The grid lines on his map became the lines of longitude

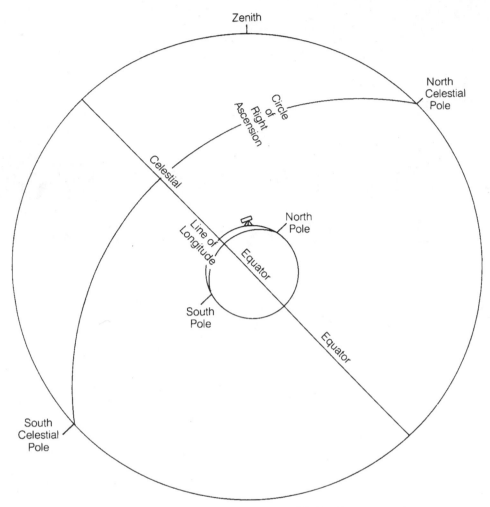

FIGURE 6-1 *The equatorial coordinate system is the extension of the Earth's own system into space.*

and latitude we use today. It took another 300 years for this type of grid to make it from the Earth to the sky.

Called the *equatorial coordinate system* (Fig. 6-1), the celestial grid takes the Earth's equator and projects it onto the dome of the sky, forming the *celestial equator*. At an angle of 23.5 degrees to the celestial equator is the *ecliptic* or apparent path the Sun follows across the background of stars in the course of a year. In the same manner, if you project the north and south poles of the Earth into space you get the *north* and *south celestial poles*. With these points of reference, the final step is to project the lines of longitude and latitude onto the dome of the sky.

The east-west position measurement given by longitude on Earth now becomes *right ascension* (RA). It is measured in hours (h), minutes (m or ′), and seconds (s or ″), starting at 00 hours, 00 minutes, and 00 seconds at the equinox and increasing as you move eastward along the celestial equator from the point of the vernal equinox. Right ascension is the difference between the object's time of transit and the time the vernal equinox transits the meridian.

The *declination* of an object is the distance, measured in degrees (°) and minutes (m or '), of the object either north or south of the celestial equator. A plus sign (+) indicates that the object is north of the equator. A minus sign (–) indicates that it is south of the equator. A declination of +56 degrees tells you the object is 56 degrees north of the celestial equator; –10 degrees puts the object 10 degrees south of the equator. The altitude of Polaris above your northern horizon is equal to your latitude. Straight overhead is what is called your latitude line. If you live at 41.5 degrees north latitude, the declination line that runs through your zenith will be 41.5 degrees.

The highest point in the sky that an object can reach is its *colatitude*. This is particularly useful when trying to locate objects with a declination far below the equator. Add your latitude to the object's declination to find the colatitude. For example, Chicago has a latitude of 42 degrees north. If you want to find out how far an object at declination –33 degrees is above the southern horizon, add 42 + (–33). That tells you the object is 9 degrees above the southern horizon. If the colatitude of the object is zero or less than zero, it will never rise above your horizon.

HORIZON COORDINATES

A somewhat simpler set of coordinates can place the observer at the center of things. Using *horizon coordinates*, the location of a celestial object is determined by its relation to the observer and the observer's horizon. This system has been used for centuries and is still used primarily for navigation.

To get the basics of the system (Fig. 6-2), go outside and face north. The point directly above your head is called the *zenith*. The point exactly opposite the zenith, or directly below you at the center of the Earth, is called the *nadir*. Directly in front of you is the *horizon*, or the line where the sky seems to join the Earth. This line travels in a great circle all the way around you. The north point on the horizon is the point directly under the north celestial pole. The line that stretches from this point on the horizon through the zenith and ends at the south point is the *meridian*. When an object crosses the meridian, it is at its highest altitude above the horizon and said to be in *transit*.

The horizon system is excellent for an observer trying to describe where an object is on the celestial sphere. Any object's position can be described by its *altitude* and its *azimuth*. Altitude is measured in degrees starting from the horizon and moving toward the zenith. An object halfway up into the sky would have an altitude of 45 degrees. If the object were directly overhead it would have an altitude of 90 degrees. Azimuth is measured in degrees from the north point on the horizon toward the east. The north point is 000 degrees, the point due east is 090 degrees, the south point is 180 degrees, and the west point is 270 degrees. An object above the horizon has the same azimuth as the point on the horizon directly below it. So a star directly over the west point on the horizon would have an azimuth of 270 degrees.

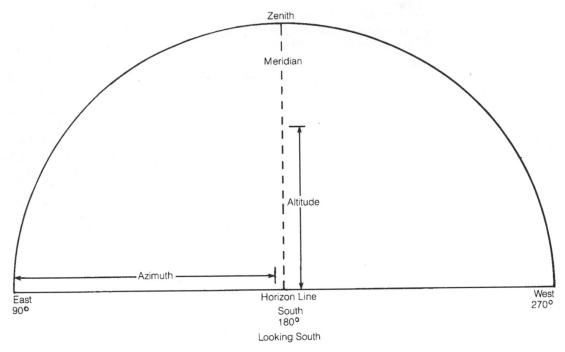

Zenith

Meridian

Altitude

Azimuth

East
90°

Horizon Line

South
180°

Looking South

West
270°

FIGURE 6-2   *The horizon coordinate system.*

TIME

We keep track of time using a basic unit of measure called the *mean solar day*. It is based on the time between transits of the Sun. In order to simplify everyday activities, the solar day is based on a Sun that moves against the background of stars at a constant rate of 24 hours between each transit. Unfortunately, this is not an accurate rate because the Earth's orbit is not circular but elliptical. As the Earth moves in its orbit around the Sun, our star appears to move at different rates and does not transit at the same time everywhere it is visible.

To lessen confusion, *local time* and standard time zones were invented. With this system, the Sun transits the meridian at noon no matter where you are. As you pass from one time zone to the next, you change your watch by one hour depending on whether you are traveling toward or away from the central timekeeping meridian, the *prime meridian*.

Astronomers selected the location for this meridian in much the same way as they selected the point on the celestial sphere marking the 00h00m00s coordinate in right ascension. They said, "This is as good a place as any; besides, there is an observatory already here so we won't have to build one." So the prime meridian, the meridian from which all time is measured on Earth, happens to run through the Royal Observatory in Greenwich, England. Local time in Greenwich is also the basis for *universal time* (UT), which is measured on a 24-hour clock and is the standard method of keeping time by both amateur and professional astronomers.

TABLE 6-1.   UT/Local Time Conversion

UT = Local Time + Correction

LOCAL TIME IN	CORRECTION
Eastern Standard	+ 5 hours
Eastern Daylight	+ 4 hours
Central Standard	+ 6 hours
Central Daylight	+ 5 hours
Mountain Standard	+ 7 hours
Mountain Daylight	+ 6 hours
Pacific Standard	+ 8 hours
Pacific Daylight	+ 7 hours

As you move west from Greenwich, you begin to fall behind UT and must add an hour for each time zone you are west of Greenwich. Table 6-1 shows you how to add to or subtract from local time to get UT. If you live in Chicago, for example, you are six time zones away from Greenwich—so you must add 6 hours to your time to get UT. This means that 1500 hours, or 3 P.M. Central Standard Time, in Chicago becomes 2100 hours UT. If an event is to begin at 0400 hours UT on June 24, subtract 6 hours to get 10:00 P.M. Central Standard Time. Remember that the date changes when you pass 0000 hours so you lose one day, making it the 23rd instead of the 24th. Also remember to adjust for daylight savings time.

Sidereal time is another kind of time used in astronomy. Simply put, sidereal time is the right ascension of objects that are on the meridian at that moment. For example, the sidereal time for midnight, July 15, is 19 hours and 30 minutes. Now you know that any object with a right ascension of 19h30m will be on the meridian at that instant. Sidereal time makes it easy to use the coordinate systems we looked at in the beginning of this chapter to actually find something in the sky. Table 6-2 gives the sidereal time of objects transiting the meridian for any hour of any day of the year.

For example, if you live in Chicago you are in the Central Time zone. To convert 6 P.M. CST to UT, first convert your local time to 24-hour time by adding 12 hours to P.M. times (leave A.M. alone). So 6 P.M. CST becomes 1800 hours. UT = local time (1800) + correction (+ 6 hours). UT = 1800 + 6. UT = 2400 (or 00).

If you know the sidereal time and the right ascension of any celestial object, you can find out if that object is favorably placed for observation. This is because stars on the equator rise about 6 hours before they transit and set about 6 hours after transit. If you subtract the right ascension of the object from the sidereal time you find its *hour angle*.

The hour angle tells you how far the object is from transit or how far it has moved since transit. If the sidereal time is smaller than the right ascension, then the object is east of the meridian. It has not yet reached the meridian and its best placement for viewing. If the sidereal time is greater than the right ascension, then the object is past the meridian and is on its way toward the horizon and setting. If the sidereal time and right ascension are equal, the object is on the meridian—and it's time to get the telescope out. You can use Table 6-2 to tell you when any object for which you know the RA will transit.

TABLE 6-2. SIDEREAL TIME

TIME	8 p.m.	9 p.m.	10 p.m.	11 p.m.	12 a.m.	1 a.m.	2 a.m.	3 a.m.	4 a.m.
DATE				RIGHT ASCENSION ON MERIDIAN					
Jan 5	03h	04h	05h	06h	07h	08h	09h	10h	11h
21	04h	05h	06h	07h	08h	09h	10h	11h	12h
Feb 5	05h	06h	07h	08h	09h	10h	11h	12h	13h
20	06h	07h	08h	09h	10h	11h	12h	13h	14h
Mar 7	07h	08h	09h	10h	11h	12h	13h	14h	15h
22	08h	09h	10h	11h	12h	13h	14h	15h	16h
Apr 6	09h	10h	11h	12h	13h	14h	15h	16h	17h
22	10h	11h	12h	13h	14h	15h	16h	17h	18h
May 7	11h	12h	13h	14h	15h	16h	17h	18h	19h
22	12h	13h	14h	15h	16h	17h	18h	19h	20h
Jun 6	13h	14h	15h	16h	17h	18h	19h	20h	21h
22	14h	15h	16h	17h	18h	19h	20h	21h	22h
Jul 7	15h	16h	17h	18h	19h	20h	21h	22h	23h
22	16h	17h	18h	19h	20h	21h	22h	23h	00h
Aug 6	17h	18h	19h	20h	21h	22h	23h	00h	01h
22	18h	19h	20h	21h	22h	23h	00h	01h	02h
Sep 6	19h	20h	21h	22h	23h	00h	01h	02h	03h
21	20h	21h	22h	23h	00h	01h	02h	03h	04h
Oct 6	21h	22h	23h	00h	01h	02h	03h	04h	05h
21	22h	23h	00h	01h	02h	03h	04h	05h	06h
Nov 6	23h	00h	01h	02h	03h	04h	05h	06h	07h
21	00h	01h	02h	03h	04h	05h	06h	07h	08h
Dec 6	01h	02h	03h	04h	05h	06h	07h	08h	09h
21	02h	03h	04h	05h	06h	07h	08h	09h	10h

ROAD MAPS TO THE STARS

Every amateur needs a good star atlas. A star atlas is nothing more than a detailed map of the sky broken down into manageable segments. Star atlases are different from the constellation guides you used earlier in two basic ways. First, they show what the constellations look like when viewed in total. A small guide such as *Olcott's Field Guide to the Skies* is excellent for the beginner, but it doesn't give you a feeling for the area of the sky that it actually covers. It's not that the little field guides are bad—it's just that if you are doing serious observing you need a more in-depth atlas.

That brings us to the second major advantage of a good star atlas—its detail. Most of the star guides show the stars down to the naked-eye limit, or magnitude 6. When you find yourself at the eyepiece of a telescope or a good pair of binoculars you are seeing stars well past that limit. Depending on the size of your telescope, you could be trying to pick your way through a field of 10th- or 11th-magnitude stars, which can sometimes be confusing. Even if you only use an accurate finderscope you are going to be able to see to magnitude 9 or better, so you need an accurate guide to what is out there. It is frustrating and a bit embarrassing to spend time at the telescope and never be quite sure of just what you are seeing. With a good star atlas you will always know just where you are looking and what you are viewing.

Star atlases come in a number of sizes and types. Some are good for locating monthly major stars and planetary data, such as Guy Ottewell's annual *Astronomical Calendar* and magazine offerings, such as the *Sky & Telescope: SkyWatch* guide. They range from simple, easy-to-use atlases like *Edmund's Mag 6 Star Atlas* to the extremely detailed *Uranometria 2000* and *Sky Atlas 2000.0*, which show stars down to magnitude 9.5 and 8.5, respectively.

Falling nearer to the Edmund atlas is probably the best star atlas for a beginner. It is the classic *Norton's Star Atlas and Reference Handbook* (the newest edition is for the year 2000). Originally published at the end of the nineteenth century, Norton's has been a staple for amateurs for over 100 years. The maps are only part of its content—it is also packed with such things as definitions of astronomical terms, information on stars and star structure, and maps of the Moon and Mars. Over a dozen maps cover the sky from the north to the south celestial pole. Although the atlas only goes down to magnitude 6, it lists many fainter stars, especially variable and double stars. Most atlases show magnitudes by varying the size of the star images; Norton's makes each magnitude a distinct and easily recognizable image, making it easy to tell the difference between a third- and fourth-magnitude star.

This discussion of magnitude representation brings to mind something that you must remember when using most star atlases. Many atlases on the market use the "lump" system of magnitude representation. In other words, they take all the stars within a certain magnitude range—say between magnitude 4.0 and 4.9—and list them as magnitude 4, using the same symbol. It's easy to understand why they do this: It would be almost impossible to distinguish little dots showing stars of magnitude 4.3 and 4.6. Longtime users are aware of this practice and do not have a problem using an atlas set up this way, but it can be confusing for beginners.

Other star atlases the beginner might consider using are *The Cambridge Star Atlas*, *Bright Star Atlas 2000.0*, *Sky Atlas 2000*, and *Uranometria 2000.0*. *The Cambridge Star Atlas*, put together by Wil Tirion, and is a guide to the entire sky for beginning and intermediate skywatchers. It has 20 charts, with stars to magnitude 6.5, and color-coded deep-sky objects.

The short *Bright Star Atlas 2000.0* is for beginners and is also written by Wil Tirion. It has 10 black-and-white maps, features 9,096 stars to magnitude 6.5, and lists deep sky wonders.

Sky Atlas 2000 was designed and plotted by Wil Tirion with Roger W. Sinnott and is considered the standard against which all other star atlases are measured. It is concise and easy to use, and is available in three versions: the field version with white stars on a black background, the desk version with black stars on a white background, and the deluxe color version with the Milky Way and deep-sky objects color coded so that there is no way to confuse a faint galaxy with a globular cluster. Tirion went to great pains to make *Sky Atlas 2000* a usable atlas, and he and Sinnott have succeeded marvelously. The atlas breaks the sky into 26 manageable sections. Each star is listed with its Greek letter designation or its Flamsteed number (numbers representing stars from west to east across a constellation) along with its common name.

Alpha Lyra, for example, is listed as Vega, Alpha Lyra, and 3 Lyra so that you know what you are looking at on the chart. It shows 81,312 single, multiple, and variable stars of magnitude 8.5 and brighter.

Uranometria 2000.0 is a two-volume set. Volume 1 covers the sky from the north celestial pole (NCP) to declination –6 degrees; volume 2 covers the sky from declination +6 degrees to the south celestial pole (SCP). Each volume contains 259 charts with a wealth of detail for the deep-sky observer—including 332,000 stars to magnitude 9.5 and 210,300 deep-sky objects.

Other, more advanced atlases are also available. The *Millennium Star Atlas* by Roger W. Sinnott and Michael A. C. Perryman is the first all-sky atlas based on the stellar positions, brightness, and distances measured by the European Space Agency's (ESA) Hipparcos satellite. The three volumes show more than a million stars to visual magnitude 11, more than 22,000 multiple stars, and over 10,000 nonstellar objects. The *Celestia 2000* brings the observer into the realm of the computer: This atlas is a CD-ROM. The disk is also from the ESA, containing stellar data from their Hipparcos and Tycho Catalogues in compressed binary format. The software interface allows users to check out the astrometric and photometric data for the million-plus stars observed by the Hipparcos satellite.

Two more atlases are the venerable *Smithsonian Astrophysical Observatory (SAO) Atlas* and the *AAVSO Star Atlas*, both of which go to magnitude 9.0. *Atlas Stellarum* is published in two volumes, one for stars from the north celestial pole to declination –25 degrees and another for stars from declination –15 degrees to the south celestial pole. It has a limiting magnitude of 14. However, these atlases are expensive and hard to find. They are also a bit hard to use: There is nothing to indicate the star you are observing without resorting to overlays and a catalog—and the same goes for trying to pinpoint the star for which you are hunting.

Consider trying these last few atlases after you have become adept at using an atlas like *Sky Atlas 2000*. Trying to use them at the beginning of your astronomical apprenticeship may only cause confusion—and needless frustration. Regardless of what atlas you decide to use, take a little time to familiarize yourself with its idiosyncrasies. Learn what symbol represents each deep-sky object and how those objects are named. One drawback of Norton's is that all the deep-sky objects—clusters, nebulae, and galaxies—carry the same symbol and are named under the old Herschelian system that divides deep-sky objects into eight subclasses. Using this system, the magnificent Double Cluster in the constellation Perseus is labeled H.VI 33 and H.VI 34; *Sky Atlas 2000* clearly labels it as the Double Cluster, with its NGC designation of NGC 869 and NGC 884 next to it.

FINDING YOUR TARGET

Picking an individual star or deep-sky object out of a star-filled sky has given more than one amateur fits. At first glance it would seem impossible to pick one's way through the vast starfields of Cygnus to locate a small open cluster of stars or a variable star near minimum. It may seem impossible—but it is not.

There are two basic ways for you to find objects in the clutter of the sky. You can dial up your target by determining the object's coordinates and adjusting your telescope's setting circles to those number. Not long ago, the only telescopes using this method were those with an equatorial mount. (Setting circles are nothing more than indices that give a numeric representation of where the telescope is pointed.) Today, however, in the age of microchips and minicomputers, even the lowly altazimuth mount can be fitted with devices to make pointing the telescope easier. A hand calculator and a couple of protractors can enable you to find faint objects using the same system with any type of mount.

We will now put our discussion of coordinate systems from the beginning of this chapter to good use. A star can be located if you know its declination and right ascension. Of those two figures, the RA is the most important to use with setting circles. To use the setting circles on an equatorial mount, set the telescope on a star for which you know the coordinates (Fig. 6-3). Adjust the RA circle so that it agrees with the published coordinates. Your telescope is now in tune with the sky. If you have a clock drive, turn it on. The setting circles will remain correct for the rest of your observing session. If you don't have a drive, don't worry. You will have to reset your circles periodically throughout the session, but, as we have seen, this is not a difficult procedure.

Using circles on an altazimuth mount, such as a dobsonian, is a bit more complicated, but only until you get used to it. Using a simple set of equations, you can convert the equatorial coordinates of any object to coordinates for the object's azimuth and altitude. On an equatorial mount the setting circles are usually built in, but with an altazimuth mount you are going to have to make your own. There are a number of ways to do this. The simplest is to use a drafting protractor for the altitude circle and scribe out an azimuth circle on the telescope ground board. With these additions and a hand calculator or computer you can easily convert equatorial coordinates to altitude and azimuth to make a setting circle system that ready works. A number of

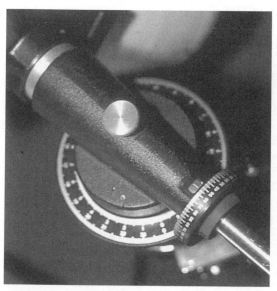

FIGURE 6-3 *Setting circles.*

recent articles describe the computer programs for these conversions; some of them are listed in the bibliography at the end of this book.

Many amateurs taken with technology say that using setting circles and computers to find objects in the sky allows you to see many more objects during an observing session. But is that what amateur astronomy is all about? Is astronomy nothing more than a numbers game in which you try to find more and fainter galaxies than the other amateurs you know? For some of us this is true. There is the amateur who preens at the fact that his computer system allows him to see over 30 deep-sky objects in just one hour. That's 30 objects in 60 minutes, which comes out to 2 minutes per object. Just 2 minutes? Isn't that giving the myriad of detail locked in the Rosette nebula the short end of the stick? Many amateurs feel that it isn't, but then again many of us feel that you should take your time with the sky and its wonders. There is no need to race through the constellations. Relax and enjoy the view.

It's not that you shouldn't believe in technology, but there is no need to take that technology and run a race with it. The microchip was developed to relieve us of the drudgery and boredom that certain things bring—in this case the computations for determining coordinates. This technology is best used when it allows you to take the time to really look at an object such as the Ring nebula or the Whirlpool galaxy, not to ring them up as points on a cosmic pinball game. All this brings us to the other way of finding objects in the night sky—star hopping. Star hopping is one way for an observer to really get to know the night sky. It can be confusing and even frustrating at times, but it is worth the effort.

The basis for star hopping is simple. You know the location of object A, which is fairly bright and obvious, and you know where object B is, which is rather faint and hard to see. What you want to do is get from A to B as simply and as easily as possible. Many amateur astronomers use this method to study variable stars. For example, start with a fairly bright star and work your way over to the variable star field you want to study. Your target doesn't have to be a variable star field, it could be a nebula or some other deep-sky object.

Star hopping allows you to become even more familiar with the night sky that crosses your telescope field. It also lets you get to know your instrument better—you learn the little foibles that hide inside a refractor or reflector telescope. It allows you to learn your limits and the limits of your equipment—and in the process helps you push those limits back a little. So let's take a few minutes and see just how star hopping works.

The first things you have to know are the field sizes of both the finderscope and eyepiece you will be using. It's really quite simple: Take your telescope out one night when you can see the Big Dipper (which is any night, because you can always see this constellation). Aim the finder at the two pointer stars (Fig. 6-4). Do they fit in the finder's field? If they do, your finder has a field of 5 degrees, because they are 5 degrees apart. If you have a lot of room left over in the finder field, check the bottom two stars in the bowl. Do they fit? If so, your finder has a field of 8 degrees, because those stars are 8 degrees apart. Most finders manufactured today have fields between 5 and 8 degrees.

Let's try using this method: We determine that the finder on our 80-mm refractor has a field of almost 8 degrees, because the bottom two bowl stars don't quite fit. Round the field off to 7 degrees; now you know your finder's field in degrees. Now scurry inside, find an old coat hanger, and grab your star atlas. Take a look at the introduction and find the scale of the atlas. If nothing like this is listed, measure the

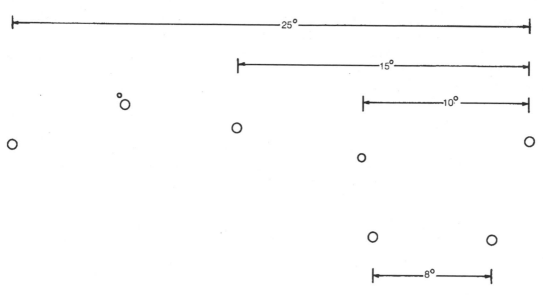

FIGURE 6-4 *The Big Dipper offers a ready-made measuring stick to find your binoculars or finderscope's field of view.*

distance between the declination lines along the side of the charts. For example, in Norton's, using a ruler, we find that 1 degree equals 3.5 mm—so the scale is 1 degree of sky equals 3.5 mm. Now take the coat hanger, snip off the straight part at the bottom, and form a circle equal to your finder's field multiplied by the scale of your atlas. For our example, this comes out to 7 times 3.5 mm or 24.5 mm. Place a bit of tape over the rough edges so you don't cut yourself. Now you have an instrument that shows what you will see through your finder. Lay this "precision" instrument on any part of your atlas. You will be able to see how much of the sky around a star will be visible to you through the finder.

Now you have to determine the field of your eyepiece, so go back outside again. This time select a star from your atlas on or near the celestial equator. A good star to use during the fall and winter months is the top star in Orion's belt, Delta (δ) Orion. In the spring and early summer use Gamma (γ) Virgo, and during the late summer and early fall use Theta (τ) Aquila. Find the star and center it in the eyepiece. Notice how it moves across the field. Now position the star so that it moves across the field, as close to the center of the field as you can. Take your stopwatch, or a watch that keeps accurate time in seconds, and time the star as it moves across the field in minutes and seconds. Do this a couple of times and take the average of your timings. Multiply the result by 15 and you will have the field of the eyepiece in minutes of arc, because it takes a star on the celestial equator 15 minutes to move through one degree. Finally, divide the final result by 60 and you have your field in degrees.

For example, if it takes your target star 5 minutes to cross the field, multiply 5 times 15, which means the field is 75 arcminutes across. Divide that by 60 because there are 60 arcminutes in one degree, and you find that your field is 1.25 degrees across. Now go back inside and twist the other end of that precision coat hanger into a circle 3.5 mm times 1.25, or about 4.3 mm across. That is the area you will see in the eyepiece—quite a bit smaller than the circle you made for the finder.

Now you are ready to star hop with the best of them. Let's try to find the Dumbbell nebula in Vulpecula using this method. Coat hangers ready? We will be using Chart 13 from Norton's, so find the proper page and turn the book upside down. (Remember, your finder is a refractor and gives you an inverted view of the sky. Turning the atlas upside down matches this view and makes it easier to follow along at the telescope.)

Pull up a chair and get to work. First let's find Beta (β) Cygri, better known as Alberio. Place the larger of your rings so that it is centered in the circle. You will notice that a star labeled Alpha in Vulpecula is at almost 12 o'clock in your field. Moving around the circle you will see a star listed as 10 Vulpecula at about 2 o'clock. This is our next target. Now move your ring so that 10 Vulpecula is centered. By the way, feel free to take notes as you go along—they will help once we get to the telescope.

With 10 Vulpecula centered, you will see our next stop, labeled 13 Vulpecula and located again at about 2 o'clock. Now move your ring to center this star. Notice that there is a group of three stars just above 13. Our final target, the Dumbbell nebula, is just above that center star. Take the smaller ring and center that on this star, labeled 14 Vulpecula. The circle covers both 14 and the Dumbbell, which is listed as M27. This way you know that if you can see 14, the Dumbbell isn't far away.

We reached an object that is beyond our naked eye by hopping from Alberio to 10 Vulpecula to 13 Vulpecula to 14 Vulpecula—with four jumps of the telescope. Now let's go outside and see if this really works.

First, find Alberto in our finder. Now switch to the main scope and take a couple of minutes to let your retina soak up the beauty of this double star. The colors are striking: The brighter of the two stars is a deep, golden yellow and the dimmer of the two a brilliant sapphire blue. The stars revolve around a common center and are about 410 light years from Earth. Now, go back to the finder; locate and center 10 Vulpecula. Take a look through the scope again. This beautiful area is on the rim of the Milky Way at one of the great rifts caused by dark matter in the arms of the galaxy. Back to the finder again and locate 13 Vulpecula. This is another nice field; 13 Vulpecula is a double star, but the two stars are too close together to be separated by anything smaller than a 6-inch telescope. If you are using a 6-inch or larger scope, by all means, enjoy the view. Take one last trip to the finder and move over to little 14 Vulpecula—and the grand finale. With 14 centered in the eyepiece, the Dumbbell nebula should be at about 1 o'clock. Take your time to find the object. As you move your eye around the field, you will see a slight blur—the nebula. As your eye becomes more adapted, you will see it better and better.

TELESCOPIC IMAGES

When you first look through your telescope, everything is going to look wonderful. The star images will be crisp, tiny points of light, star clusters will fill the field, and the Moon will look unbelievably beautiful. But then, sooner or later, you are going to consult a star chart or a map of the Moon and the confusion will start. Nothing will

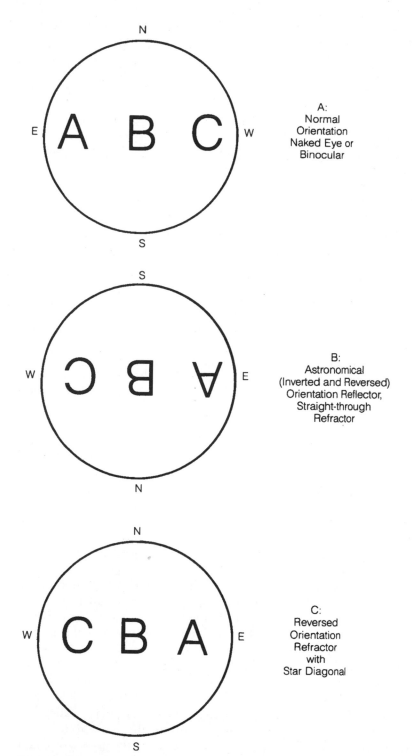

A:
Normal
Orientation
Naked Eye or
Binocular

B:
Astronomical
(Inverted and Reversed)
Orientation Reflector,
Straight-through
Refractor

C:
Reversed
Orientation
Refractor
with
Star Diagonal

FIGURE 6-5   *Field orientations.*

match. The stars all seem to be in the wrong places. What you see on the chart is nothing like what you see through the eyepiece. Is something wrong?

The simple answer is no. When you look at an object in the sky with your naked eye or a pair of binoculars, the orientation is correct. Look at an object through a telescope, however, and everything is upside down (Fig. 6-5). The simple explanation for this confusion is that most astronomical telescopes invert the image you see. Because the direction of the light making up the image is changed in some way by the telescope's objective, either reflected by a mirror or refracted by a lens, the final image reaching your eye is flipped end over end. Hold a magnifying glass up to view a lamp across the room and you will see the same effect. The image as it enters the lens is right side up, but as you move your eye back from the lens to the point of focus the image flips over. The bottom has become the top, the top the bottom. This flipped image is the normal way of seeing the universe through a telescope.

Problems start when we add surfaces from which our image will bounce off. When you put a star diagonal into a refractor, the image changes again. You still have the erect image but you have changed something else—the image is now reversed. Left becomes right and right becomes left. So now your image is erect but reversed. This can cause a great deal of confusion when you are trying to read a star chart at the telescope. You can always turn a chart upside down to conform with the telescope's inverted image, but there is no way you can move a chart around to make it conform to an inverted and reversed image.

An image will reverse its orientation each time its direction is changed by an optical surface. The number of these directional changes is the key to what you finally see. Even numbers of changes keep the image orientation normal—left to left, right to right. A newtonian reflector gives the image two directional changes—one when it reflects off the primary mirror and another when the image reflects off the secondary—so the image is not reversed.

A refractor changes the direction of the image by bending it as it goes through each lens in the system. Remember that all astronomical refractors have at least two elements in their lens system. Two elements mean two changes of direction, and the image is properly oriented for an astronomical instrument. Top and bottom are inverted, but left and right are okay. Add a star diagonal and you are adding a third optical surface that will change the image direction. Odd numbers of direction changes are what causes image reversal. A tri-schiefspiegler telescope uses three mirrors to bring the image to the eyepiece, three changes of direction, making a reversed image. A Herschelian telescope reflects the image directly to the eyepiece with no secondary mirror, resulting in one reflection, one direction change, and a reversed image.

You can remedy the problem of using a star chart with a reversed-field instrument in a number of ways. You can turn the chart over and shine a light through the back of the chart. If you use charts that are in book form, such as *Uranometria 2000.0*, you will have to copy the desired chart by tracing or photocopy. This is fine, but it can be a bit cumbersome to hold a chart and a flashlight, and still look through an eyepiece.

An easier solution is to use an *amici diagonal*. This diagonal adds an additional surface so that a refractor has an even number of image direction changes—eliminating the problem of image reversal. There is some loss of brightness due to the extra surface, but many observers think the proper orientation is worth the price. You could take this one step further. There are systems that correct for all of the image changes by putting top on top, bottom on bottom, left to left, and right to right. But these so-called erecting systems waste too much of the light entering the telescope.

Now that we have the facts straight about image orientation, we're ready for our first night observing. But there is one more thing we have to tackle before going to the backward: We have to find celestial direction in the eyepiece.

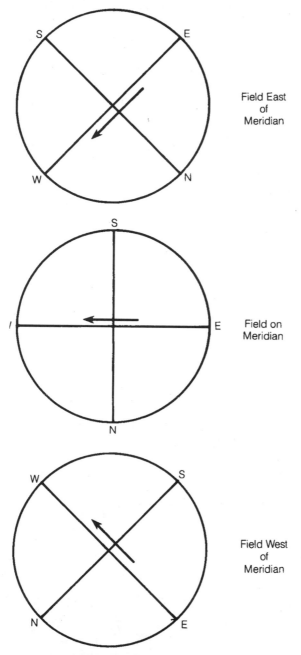

Field East
of
Meridian

Field on
Meridian

Field West
of
Meridian

FIGURE 6-6 *Directions in the field of view as it moves across the sky.*

Stars appear to rise in the east, transit the meridian to the south and set in the west (Fig. 6-6). This seems very simple to comprehend, but remember the normal telescope field is inverted and sometimes reversed. Directions can become very confusing if you aren't prepared. The night sky is a very big place—and it's easy to get lost.

Remember our inverted field? Let's give those tops and bottoms some concrete directions. The top of an astronomical telescope field is always south and the bottom is north. This may seem confusing, but there are a couple of tricks you can use to get your bearings. You can find celestial north by giving the telescope a nudge toward Polaris, the North star. The direction from which stars enter the field is north. In a nonreversed field, east is left and west is right, but in a reversed field, east is right and west is left. You can always find east by simply watching which way stars enter the field when the telescope is stationary. It's not a bad idea to check your directions occasionally during an observing session. When you are using a chart at the telescope, make sure that you orient it so that the north on the chart corresponds to north in the eyepiece. If you do this, everything else will fall nicely into place and you won't get lost.

7

YOUR FIRST
NIGHT OUT

A journey of a thousand miles begins with a single step.

CHINESE PROVERB

It has finally arrived: The darkness you have been waiting for so patiently is finally descending. You have taken the time to set up your telescope in a protected area of the yard. You've spent the last hour or so checking and rechecking the telescope's collimation. You sit back on your lawn chair and watch with new wonder as the stars begin to glow from their heights. You also frown at the broadening glow of the sodium vapor lamps that suddenly seem to surround you in what you thought was a light-secure spot.

As the twilight fades you are able to pick out the outlines of constellations that only a few months earlier were unknown to you. The bright point of light that hovers over your neighbor's roof is Jupiter. Nothing else gets that bright, that high in the sky. Almost overhead, you can discern the outline of Andromeda, the constellation that holds M31, the Great nebula. Falling toward the western horizon, but still easily visible, is Lyra with Epsilon, the famous "double double," and the beautiful Ring nebula, M57. There is so much to see. So where do you start?

Start with the atmosphere. It lies between you and your goal of the stars—and it is critical to what you will observe and how you will observe it.

THE ATMOSPHERE

We don't normally think of the atmosphere as dynamic. It is merely something we all take for granted. But take a good hard look. There is more going on than just clouds, rain, and snow. On a hot summer day, take a look down your street. The air just above the street seems to be alive, shimmering and shifting like a living thing.

FIGURE 7-1 *This photo of the Blue Ridge Mountains shows you what happens as you look farther into the atmosphere—haze and a thick atmosphere obscuring the view.*

Heat rising from rooftops in the cold of winter creates the same movement. The air is heated by either the hot street or warm roof and rises into the atmosphere. If you try looking through that rising hot air with any size telescope, you won't see much of anything.

The atmosphere can be your greatest help in the pursuit of astronomical wonder, but it can also be your worst enemy (Fig. 7-1). You have to know when to sit and work with your helper, and when to sit back and wait for its mood to change. From the standpoint of the astronomer, the Greeks had it figured out. The heavens and the weather were the domain of the god Zeus, the cloud gatherer. He would let the rains fall at his pleasure. He would decide when the clouds would obscure the stars and Mount Olympus from the prying view of man. Now we have meteorologists, weather maps, and computers. They may be more accurate, but the Greek version was more colorful.

When clouds fill the sky, it's easy to see that observing is out for the evening. It is much harder to evaluate what seems to be a perfectly clear sky. What is needed is a way to measure the quality of the atmosphere. There are, in fact, two measures of sky quality.

TRANSPARENCY

Transparency is a measure of how bright the stars appear. Another way of defining transparency is the clarity of the sky. On a good night, an observer outside the city may see stars down to magnitude 6 while another observer in the city may only see stars to magnitude 5. The difference of only one magnitude seems insignificant, but over half of the 6,000 or so naked-eye stars fall below the fifth magnitude. It stands to reason that if you can see fainter stars with your unaided eye, you can see fainter stars with the help of your telescope.

There are stories of observers pushing their instruments past the limits set for them on paper. Some observers, under perfectly transparent mountain skies, can push a 60-mm refractor past its limiting magnitude of 10.9 to magnitude 12, a startling feat that, unfortunately, is not likely to happen under skies in a large city. The lack of transparency in city skies affects the limiting magnitude of telescopes as well as the eye. Limiting magnitude may not be a hard-and-fast rule, but it is one that changes with the prevailing conditions.

Any haze, smoke, smog, or dust particles will work to reduce transparency. Another factor is the amount of light "splashing" up into the sky. The light hits the particles suspended in the atmosphere and lights them up, thus causing a kind of artificial "aurora." This is of particular concern to observers who live in and near large cities. Many skies are often plagued by city lights bouncing off a combination of haze, smog, and dust. It is sometimes so bad that it appears to be another horizon—a "light horizon." You can't see anything below it, no matter how good the overall transparency.

Smog, dust, and haze hang in the atmosphere, and there is nothing you, as an observer, can do about them. There are ways to overcome the problems presented by those pesky streetlights, however. First, if you try, you can almost always find a spot sheltered from their glow. A tap tree or bush, a convenient garage, or a tall fence can serve to block the entry of extraneous light. If it's impossible to find such blocks, you can always go the route of Sir William Herschel and drape a hood over your head while at the eyepiece. It does work: The amount of light cut down by a good, dark cloth draped over your head is phenomenal. If you don't want to go to such extremes, just cup your hand around your eye while you look through the eyepiece. This will block most of the light from the street and will improve the telescopic image immensely.

There is, however, one source of light that it is impossible to overcome: the Moon. When the Moon is full, it has a magnitude of {nd}11 and its presence in the sky washes out any faint objects such as galaxies, nebulae, or faint comets. With even a first- or last-quarter Moon, the transparency suffers to such an extent that observing deep-sky objects becomes futile. Added to the lights of a large city, the presence of the Moon will probably limit your observing to the Moon itself, planets, and bright double stars.

SEEING

The other measure of sky quality is called *seeing*. This refers to the steadiness of the atmosphere you are looking through and its resultant effects on the actual telescopic image. Seeing is affected primarily by heat and turbulence in the atmosphere. Hot air rising from a street or roof causes a telescopic image to boil, rendering it useless. The boiling causes the image to become blotted and deformed until it is impossible to distinguish anything but a featureless blob through the eyepiece. Telescopic images also tend to look jittery under conditions of poor seeing. There may be little if any deformation, but the image under observation looks as if it is shimmering. The naked-eye counterpart of the effects of poor seeing is what we see as the twinkling of the stars.

Turbulence tends to be of two types. High-altitude winds can push layers of hot air, making seeing that much harder to predict. Unfortunately there is nothing you can do about this problem. Turbulence closer to the ground, say up to 1,000 yards off the ground, can also cause these bubbles of warm air to move around and ruin a carefully planned observing session. One good thing about this lower-level turbulence is that it usually dissipates by about midnight, so you can easily wait it out. Winds and turbulence at ground level can also cause the telescope to vibrate and the image to deteriorate. You can take actions to protect your telescope (and yourself) from these ground-level winds by finding an observing spot with a windbreak of some kind.

Finally, turbulence and heated air can combine to swirl around inside the telescope tube—creating poor viewing. Of all the potential observing problems this is the easiest to remedy. Set your instrument outside well before you will actually begin using it. This gives the air temperature inside the tube time to stabilize to the outside temperature. This turn to ambient temperature will definitely improve seeing.

MEASURING SEEING AND TRANSPARENCY

The early twentieth-century, Turkish-born French astronomer E. M. Antoniadi devoted his observational life to the study of the planets. He understood the importance of knowing under what conditions a given observation was made. In the later years of the nineteenth century, Antoniadi, working with the great French astronomer Camille Flammarion, discovered faint, elusive markings on the planet Saturn. This caused quite a stir at the time, and the two were accused of fraud in some circles. It was not until the great and keen-eyed early twentieth-century American astronomer Edward E. Barnard saw the same markings that the French astronomers were vindicated.

In the 1920s, Antoniadi proposed that the so-called "canals" on Mars (a rumor started by Percival Lowell earlier, who believed the markings were evidence of intelligent life—but that's another story) were only illusions. He believed the eye linked details on the planet's surface, creating what looked like a meaningful pattern. In each case, the observations depended on stable seeing conditions existing at the time of the observation. From his experiences Antoniadi developed the following scale for seeing conditions:

I. Perfect seeing, without a quiver
II. Slight undulations, with numerous moments of calm lasting several seconds
III. Moderate seeing, with large tremors
IV. Poor seeing, with constant troublesome undulations
V. Very bad seeing

Although this scale was developed primarily for use by planetary observers, it is very adaptable. The Webb Society, a worldwide group of deep-sky object observers, uses this scale to keep its records. The Association of Lunar and Planetary Observers (ALPO) uses a system that measures seeing on a scale from 0 (worst) to 10 (best). This method is a bit more subjective than Antoniadi's and depends a great deal on the individual observer's experience.

Measuring transparency, on the other hand, is a bit easier and more standardized. Most observers rate transparency by the faintest star they can see with the naked eye. The scale runs from 0 (nothing but very bright stars, like Sirius, visible) to 6 (easily seen sixth-magnitude stars). Like everything else in astronomy and life, the more experience you gain, the easier it is to measure transparency and seeing.

OTHER ATMOSPHERIC TRICKS

Probably the biggest trick the atmosphere plays on the observer is that conditions giving good transparency and seeing do not occur at the same time. For the most part, a steady atmosphere with good seeing will tend to be a bit hazy. The haze acts to calm the atmosphere and improve the amount of detail an observer will see. On the flip side, a really transparent sky can set the stars twinkling like crazy, with the telescopic images blobbing out rapidly and regularly.

The best seeing takes place just after a low-pressure system, when the barometer is on the rise. High winds tend to make a sky more transparent as they push warm, hazy air away from an area like a city, but then the observer has to contend with the resultant loss of definition and resolving power caused by the increase in turbulence.

You can, and should, try to fit your observing to the prevailing sky conditions. On nights with transparent but unsteady seeing, try to limit your observations to faint, extended objects that do not call for detailed examination. Galaxies, gaseous nebulae, and planetary nebulae fall into this category. Because they have little surface detail, they are not affected by a lack of steadiness and benefit from the greater clarity brought by high transparency.

On nights when the steadiness of the atmosphere is great, study the Moon and planets with high magnification. Double stars are also good things to study during times of high stability because of the high magnification usually required to split them. If you find yourself blessed with the rare combination of high transparency and high steadiness, try to contain yourself. Now is the time to do all of the aforementioned items and split some of those very close double stars.

All things considered, the best part of the sky to observe runs from about 45 degrees above the horizon to zenith, the point directly over your head. Below 45 degrees, the observed object's light has to pass through more atmosphere to reach your eyes. Here light is subject to what is called *atmospheric extinction*.

Atmospheric extinction can cause problems when you are trying to find a faint deep-sky object or to estimate the magnitude of variable stars. A star loses some of its brightness when its light is absorbed by the atmosphere, making it difficult to accurately compare the magnitudes of a variable star and its comparison stars. This also makes extended objects such as galaxies and diffuse planetary nebulae appear dimmer than they actually are and therefore more difficult to locate and study. Table 7-1 indicates the magnitude loss for various angles above the horizon.

TABLE 7-1. ATMOSPHERIC EXTINCTION

DEGREES ABOVE HORIZON	MAGNITUDE LOSS	DEGREES ABOVE HORIZON	MAGNITUDE LOSS
43	0.1	13	0.8
32	0.2	11	0.9
26	0.3	10	1.0
21	0.4	6	1.5
19	0.5	4	2.0
17	0.6	2	2.5
15	0.7	1	3.0

THE ART OF SEEING

Sherlock Holmes often admonished his friend and colleague Dr. Watson about the latter's lack of observational powers. When confronted with a seemingly simple piece of evidence, Holmes would purse his lips and shake his head: "You see, but you do not observe." To Holmes's trained eye the key details were obvious, but Watson saw nothing. Observational astronomy is very similar. The trained eye sees a wealth of detail, but most people are left scratching their heads and wondering what they are really seeing.

The ability to see minute detail in the swirling clouds of Jupiter, to split very close double stars, or to discern faint galaxies must be learned. Sir William Herschel went so far as to warn the new observer that he must not expect to "see at first sight." To Herschel seeing was an art—an art that had to be worked at constantly to be improved. How do you go about teaching yourself to see?

Observational astronomy is a little like weight lifting. You cannot expect to bench-press 350 pounds on your first trip to the gym. You have to develop your muscles before you lift anything of consequence. You have to develop your ability to see just as you would work a muscle. Working at the telescope on a regular basis is the easiest way to improve your ability to see. The more you use your ability to discern detail, the stronger that ability becomes.

The first thing you must learn is patience. You can spend a lot of time at the eyepiece waiting to see detail in Jupiter's clouds. The atmosphere is unsteady, and moments of really good seeing are rare and fleeting. They are there, but you have to be patient.

Looking for detail is also a bit more involved than popping in an eyepiece and racking the image into focus. Using Jupiter as an example, begin with a low magnification. Study the surface as closely as you can without straining. As you observe, you will notice that the image may be fuzzy for 5 or 10 seconds, then become razor sharp for a second or so, then retreat back to fuzziness. That's our friend the atmosphere throwing in its two cents' worth.

Try varying the magnification from time to time. Jump from your 26-mm lens to your 12-mm, and then use the 12-mm with your 2× barlow. The detail is there, and it will reveal itself to you slowly.

The eye works like a camera. It tends to record the detail over a long period of time. Remember that we mentioned that the retina can selectively filter what it's viewing. This ability comes into play here. The retina initially passes along all the images it takes in until it sees little flecks of detail. It then starts to filter out those fuzzy images and passes along only the sharp images that show the detail. In this respect the eye is much better than a camera. The camera only records a single moment in time, but the eye records a large number of images as a single moment. You won't remember that you stared at Jupiter for 15 minutes as it floated fuzzy and blurred in the field of view. You will remember the slight wisps of white that floated on the southern equatorial belt, even though you saw them for only an instant.

Now that you are a patient observer, it's time to start talking to yourself. Get in the habit of asking yourself about the object you are viewing. If you are looking at the Moon or a planet, study it under different magnifications. Take notice of what the varying magnifications do to the image. Which lens makes it easier to see, which more difficult? Does a higher magnification bring out more detail, or does the detail seem to get lost in the background of the scene? Once you have discovered a signifi-cant bit of detail with a high magnification, can you find it easily with a lower mag-nification? Ask yourself to describe, out loud, what you are viewing. For some reason, this sets what you see in your mind. If you are looking at a deep-sky object such as an open cluster, ask yourself, What is its shape? Where is it brightest, dimmest? Is there any trace of nebulosity visible? You get the idea. By looking for specific details in the object you are observing, you will gradually see more and bet-ter fine points. By talking to yourself, you will become a better observer.

You can train your eye in other aspects of telescopic seeing. One important abili-ty is to be able to judge what is referred to as *angular measurement.* This is useful in estimating distances and sizes. Whether it is an estimate of the diameter of a distant galaxy, the size of a feature near the Great Red Spot on Jupiter, or the separation of a double star, it is important to have a working grasp of the way these measurements actually look in a telescope. Angular measurements are made in arcminutes and arc-seconds. Texts on astronomy define an arcminute as $\frac{1}{60}$ degree and an arcsecond as $\frac{1}{3600}$ degree. Fine, but what do these things look like?

The best way to develop your ability to judge angular measurement is to look at a number of objects of known size, including double stars with known separations. Table 7-2 lists a number of objects and their angular size with which you can prac-

TABLE 7-2.   ANGULAR SIZE

MEASURED BY YOUR HAND HELD AT ARM'S LENGTH:

Thumb to forefinger	20 degrees
Forefinger to little finger	15 degrees
Across your knuckles	10 degrees
Your index fingernail	1 degree

MEASURED BY SKY OBJECTS:

The entire Big Dipper	25 degrees
Distance between the top stars of the bowl	10 degrees
Distance between the pointer stars	5 degrees
Diameter of the full Moon	30 arcminutes
Mizar to Alcor (at the bend of the handle in the Big Dipper)	12 arcminutes
Epsilon-1 Lyrae to Epsilon-2 Lyrae	3.5 arcminutes

tice. It would be best to use the same eyepiece for the initial observation of each object. For instance, start with a 25-mm. After a bit of practice, you will be able to transfer the apparently abstract concepts of arcminute and arcsecond to the universe you see through the eyepiece.

As soon as you look up at the stars, you will become aware that not all stars shine with the same intensity and brightness. As we saw in Chapter 2, the brightness of the

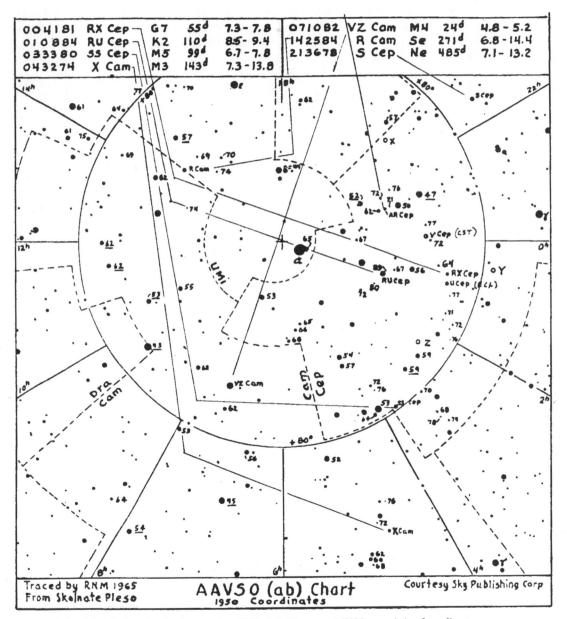

FIGURE 7-2   *AAVSO chart for the circumpolar AR Cephei. (Courtesy AAVSO, permission from director Janet Mattei)*

stars is set on a numerical scale ranging from the Sun at magnitude −26 to the faintest stars seen by the largest telescopes at about magnitude 25. Each step on the magnitude scale is 2.5 times brighter than the one that follows it. Accurate estimates of the brightness of an object are important when examining such things as variable stars and newly discovered comets. The best way to train the eye to estimate magnitude is, again, simply by persistent observation. To facilitate the development of this ability, select an area of the sky for which you have a magnitude chart. Those supplied by AAVSO for variable star study are excellent for this purpose (Fig. 7-2). These charts come with an inverted field and are ready for consultation right at the telescope. They also list the magnitudes without decimal points—magnitude 7.4 is listed as 74 and so on—so you won't confuse the point with stars. Begin with a 20- or 25-mm eyepiece, and locate the area on the chart. Examine the area carefully, and put a little checkmark by each star you can easily see. Notice how the brightness of the stars relate to each other. Use a slightly higher-magnification eyepiece, and you will see the effect increased magnification has on brightness. Can you see fainter stars? Now put the chart aside. Can you pick out the star at magnitude 6.6? How about 9.3? After a while, your ability to estimate magnitudes will be second nature.

PROTECTING YOUR VISION—AND YOURSELF

Once you have adapted your eyes to the darkness, you have to guard against any sort of bright light. If you consult a chart, use a flashlight with a red shield over the lens. This does not destroy the visual acuity you have built up because the visual purple in your eye will not react to red light. Photographers see the same phenomenon in the darkroom. Most darkroom safelights are built around a red-filtered light source. Black-and-white film and most photographic paper, like the eye, is not as sensitive to this type of light. A simple way to produce a light for chart reading is to remove the reflector from behind the flashlight bulb and then paint the bulb itself with red nail polish. The resulting light is diffused but quite adequate to check a chart or make a notation.

You also have to keep your eyes healthy and your visual acuity high. Make sure that your eyes get an adequate blood supply while observing. Alcohol and nicotine are two drugs that cut down the flow of blood to the eyes. They should be avoided, especially before and during an observing session. Blood pressure and diabetes are other conditions that can impair vision, so keep them in check.

Other factors that enter into your own "tranquillity of the retina" include the comfort of your observing position. Sitting is the best posture, but unless you are using a refractor with a zenith prism, this is difficult. Try not to tense your body while you work at the eyepiece. It is helpful to keep both eyes open while observing. This is easy if you are in a completely dark environment, but if you are observing from the city it is hard to keep distractions from gaining the sight of the unused eye. The easiest way to avoid this problem is to use an eyepatch. You can find them at many drugstores for about $1.50, a very worthwhile investment.

Next, you have to keep yourself warm. As you observe on a cool autumn evening, your body heat melts away and your circulatory system has to work harder to keep your temperature stable. This is even more of a problem during the winter. The colder you are, the less blood reaches the eye, and your vision suffers. It is not a bad idea to invest in a good set of thermal underwear. Dress in layers of clothing, because this allows layers of warm air to further insulate your body. A hat or watch cap will help retain most of the heat lost from the top of your head and add to your overall comfort. A thermos of tea is a blessing during a session with temperatures anywhere below 40 degrees, as long as you keep it away from your charts. A thermos seems to know it must spill on anything important—probably some undiscovered law of physics.

You will find that the amount of detail you see decreases over the length of the observing session. This is because your eyes get tired after being used over an observing session—and they tend to rebel. You may suddenly find it difficult to keep an object in focus, and the strain may give you a headache. The eyes ignore any image that remains in the same place on the retina for any length of time. To compensate for this, the brain orders the eye to look harder for the image. Eyestrain is the result. To overcome this problem, all you have to do is keep the eye moving around the field under observation. It is a good idea to give your eyes periodic rest breaks. One method is to switch eyes, making sure that each gets a chance to rest in total darkness under your eyepatch.

GETTING TECHNICAL

Although this book is meant to help the new observer, there are some things you will see in magazines or on other amateurs' telescopes that will entice you. Some are best looked at only after you get the feel of the universe and your telescope. Others should be approached after you are certain of your interest in astronomy. Many people jump right into the major systems, hoping to see images like those taken by the Voyager spacecraft—if they just get the right equipment. But such images are far beyond the capabilities of amateur, or even professional, telescopes.

One adventure amateur astronomers often attempt is astrophotography. But before you jump into researching f-stops and film speed, be sure to ask yourself some important questions. First, what type of imaging are you interested in doing? Although photographs of planets and deep-sky objects can be taken with certain systems, you will need customized telescopes and cameras for both sets of objects to get the best images. Second, how much are we willing to pay for such a setup, including telescope, camera, and a great deal of film? Finally, do you have the patience to practice taking long-term exposures of astronomical objects?

One state-of-the-art assemblage for astrophotography is called a CCD (charge-coupled device) system, a camera{nd}telescope system used mostly by amateurs for deep-sky and planetary objects. The system can give you a wonderful look at the universe—or make you tear your hair in frustration.

A CCD is an array of light-sensitive picture elements, or pixels, each measuring 5 to 25 microns across. The camera's lens focuses the scene onto this pixel array. Just as the resolution of conventional photographic film is related to the size of the grain, the resolution of a CCD chip is measured by the number of pixels in the array. A digital still camera intended mainly for nonprofessional use has an array of, typically, 640 by 480 pixels; a top-of-the-line professional camera would have an array of millions of pixels. Charge-coupled devices, which were developed in the early 1970s, are the most commonly used light sensors and are also incorporated in such products as video cameras, facsimile machines and desktop scanners. CCD-based cameras make it possible to capture images that can be instantly transmitted, for example, from a photojournalist in the field or from a reconnaissance satellite in space.

Simply put, in this camera-telescope system, the camera is digital—literally a light-sensing array that is like having reusable "film." The CCDs (the light sensors) gather the light much better and faster than the standard telescope-mounted "optical" camera. The frustration usually comes if you don't match the major components of the imaging system: the camera, telescope, mounting—and even the observer.

The CCD produces what are known as pixels, areas where light is gathered by the instrument. A pixel is similar to the squares you see on some spacecraft images and in close-ups of newspaper photographs—and the combination of these pixel squares equals an image. Contrary to what most people think based on a regular telescope setup, in a CCD system, a larger aperture is not necessarily important. If the observer makes a longer exposure with a smaller aperture, the images are often of the same detailed quality as those made with a larger instrument. This is one of the major draws of the CCD system.

The most important part of picking a CCD system is choosing the right camera. It is important to match the telescope's focal length to the CCD's pixel size, which determines how much sky each pixel sees. If the match is not good, the resulting image will be a strange—and undefined—pattern of small squares. Although you can enhance such an image on the computer by a process called resampling, it's often more useful to start out with a more compatible camera-telescope system at the start.

Picking out good optics is also important. If the telescope has any defects, such as a halo around the stars in a certain eyepiece or the telescope's inability to crisply focus, the CCD magnifies the imperfection. In other words, for every detail you hope to enhance in a star field or planet image the defects will also be enhanced.

The system's focuser is important, too. You need to change the focus during the course of your observing time because temperature differences can tweak the focus ever so slightly. And because CCD cameras are picky about focus, for better images, you need a prime focuser.

As with all telescopes, the mount also is important to the CCD image. Even the smallest guiding error, or jarring and bucking of a rough clock drive, can eliminate all the detail you wanted from the system. Shorter time exposures are not as much of a problem as longer exposures.

Finally, you, as the observer, become important to your CCD system—or any astrophotography system. You need to guide the system for the best images, and the only way to do this is through experience. If you do take on a CCD system, practice. Know how your telescope works, how to spot an image—and what produces the best images. Write down your experiences and keep a good log of your attempts at taking images. For more information and help on CCDs, there are many good sites on the Internet, books, and even a magazine, *CCD Astronomy* (published by Sky Publishing) that will explain more to you about this exciting new world only recently opened to the amateur astronomer.

Why CCDs? As the editor of *Sky & Telescope*, Leif J. Robinson, once pointed out,

> Hundreds of amateurs have telescopes 0.2-m to 0.5-m in aperture that are equipped with CCDs and other high-sensitivity, low-noise accessories. They also have the computing power to carry out thorough data analysis. . . . Knowledgeable, equipment-rich amateurs could make excellent collaborators with professionals. In fact, many amateurs may desire to work directly with a professional, one-on-one. Such interaction is now exceedingly easy thanks to modern communications . .

Now we need both sides to come together—professional and amateur—to expand the information gathering CCDs provide.

TUNING IN

There is another area where amateurs are beginning to make a difference—the field of radio astronomy. For someone interested in astronomy and radio electronics, such an endeavor can be rewarding. This is a relatively new field, so don't be surprised if you don't find many people who understand your endeavor. But the use of small radio telescopes is increasing—especially with access to more satellite dish technology.

The best way to determine if this is the way you want your astronomy to go, contact the amateur group that deals with radio astronomy: the Society of Amateur Radio Astronomers (SARA, at *http://www.bambi.net/sara.html*). They claim that about 65 percent of our current knowledge of the universe has come from radio astronomy, including the discovery of black holes, quasars, and pulsars. These may be the realm of professional radio observatories, but amateurs can help, too—by looking at broad areas of the sky for long periods of time, something the larger observatories don't do.

Another way to determine your interest is to contact amateur radio operators (often called ham operators), who often have skills and equipment that slip right into radio astronomy. There are many ham radio clubs across the United States (and the world). See if you can join a club, and/or get a ham radio operator's license. It will get you closer to radio astronomy—and may lead to two great hobbies.

What radio astronomy equipment do most amateur radio astronomers use? In general, the amateur radio telescope system includes a good antenna system, a sensitive, stable, low-noise receiver, and various output devices. In most cases, the output is in the form of a voltmeter, a data-logging computer, or a strip chart.

Amateurs can contribute in many ways. In particular, they hope to discover a new radio source. Table 7-3 explains in general some of the areas in which amateurs can often contribute (most of these can be conducted with either imaging or non-imaging techniques). But there are restrictions. For example, you would need more advanced equipment to study pulsars, such as a very high-frequency (VHF)/ultra-high frequency (UHF) receiver and a very good antenna. For studies with the Search for Extraterrestrial Intelligence (SETI), you would need a much larger and

TABLE 7-3.

Monitoring	Expertise	Equipment Needed
Jupiter noise storms	little	shortwave and dipole antennae
Meteor studies	little	VHF receiver and antennae
VLF solar flare monitoring	little	VLF receiver and antennae
High-frequency/very high-frequency sources	more advanced	high- frequency/ultrahigh frequency receiver and antennae

complex system. In addition, you can also work on radio astronomy based on the data already out there, because much of it is accessible over the Internet.

You may even end up joining a research project already in progress. For example, the Woodbury Research Project's purpose was to retrofit a 100-foot satellite dish to operate as a radio telescope. The project at Georgia Tech included converting two abandoned AT&T satellite dishes into radio telescopes. Another project is at the Haystack Observatory at the University of Massachusetts. The Small Radio Telescope (SRT) is being developed to observe in the L-band (1.42 GHz)—a 10-foot-diameter satellite television dish mounted on a fully motorized Az-El mount. It's a "prototype" for an inexpensive radio astronomy kit that the researchers hope will eventually be used to introduce students and amateur astronomers to the field of radio astronomy.

GREAT EXPECTATIONS

No matter what you chose to do—from using a regular telescope to CCDs to radio telescopes, remember you have to start at the beginning. It takes time, but the rewards—and the wonderful people you meet along the way—are worth the effort. But overall, always remember to keep your expectations under control.

When you bought the telescope at the garage sale, the guy said he was selling it because he couldn't see anything over the glare of city lights. You contend with the same lights, however, and you can see plenty of stars. Could it be something more than lights? If you could talk to him tonight, with the telescope all set up and ready, he would probably tell you that there must be something wrong with the telescope. It shows the Moon line, but when you look at Mars you can't see the dark lines. You can't even see Mercury's surface, Jupiter doesn't show the multicolored bands across its face, and Saturn's rings just don't look right (Fig. 7-3 and 7-4).

The person who sold you the telescope suffered from one of the most common afflictions known to the amateur astronomer: He expected too much from a small instrument. To give you a normal example, at a recent observation session held by a local astronomy club, a little girl squinted through the eyepiece of a 60-mm refractor. She chewed her lower lip as she looked at Jupiter through a low-power eyepiece. Climbing down she sighed and shook her head.

FIGURE 7-3   *The planet Jupiter from the Voyager spacecraft. Photos like this may lead many people to expect more from their telescopes than the instruments can deliver. (NASA)*

"What's wrong?" the amateur astronomer who owned the telescope asked. "Didn't you see Jupiter and its moons?"

She shrugged. "I guess so, but it doesn't look like the stuff on TV." Our little friend is suffering from the same problem as the guy who sold you the telescope: reality not meeting expectations.

We have been spoiled by our advanced technology. The views that the Voyagers 1 and 2 sent back as they whizzed through the outer solar system have given us a preconceived image of Jupiter and Saturn. Even the short-lived Pathfinder on the planet Mars and the Mars Global Surveyor showed us fantastic views of the red planet. When we approach a telescope, whether it's a 60-mm refractor or a 17.5-inch dobsonian, we feel that those are the images we should see. Thus, the reality of the eyepiece cannot be anything but disappointing.

FIGURE 7-4   *This mosaic of the planet Mercury was taken by the Mariner 10 spacecraft in March 1974; you will never see this image in your telescope. (NASA)*

So the very first thing you must do as you approach the telescope for the first time is clear your mind of any preconceptions. If you don't, you will be disappointed regardless of what you see. Approach with an open mind, and the wonders displayed before you will reward your patience and imagination tenfold.

So what do you do now? Get your telescope ready and follow along as we begin our journey through the universe.

8

OBSERVING THE MOON

. . . we shall find the Moon a wonderful object of study.

REV. T. W. WEBB

The Moon is a unique object for the amateur astronomer to study. It is visible somewhere in the sky almost every night. Outshining everything in the sky but the Sun, it is easy to find and identify. Above all, it is the closest celestial object to Earth with the exception of meteors. Viewed with the smallest telescope, or even a pair of binoculars, the Moon yields more detail than can be seen on any other object in the sky. With all this going for it, it is understandable that the Moon is usually the first thing amateurs view with their telescopes.

For 360 years, the Moon was a special place for humanity in general and amateur astronomers in particular. All that changed on July 20, 1969. For some reason, as soon as Neil Armstrong placed his foot on the surface of Mare Tranquillitatis (or "Tranquillity Base"), the Moon ceased to be an object of mystery and became just another place on humanity's vacation list. It went from being The Moon to being the moon. The missions of the Surveyor, Lunar Orbiter, and Apollo programs mapped the lunar surface so well that professional astronomers lost almost all interest in looking at the Moon with Earth-based instruments (Fig. 8-1).

This filtered down to the amateur ranks almost immediately. Prior to the age of the Lunar Orbiter and Apollo, *Sky & Telescope*, a magazine for American amateur astronomy, carried a monthly column for amateur lunar observers. Almost every month articles on specific features on the Moon could be enjoyed by enthusiastic amateurs. Then, suddenly, they stopped. Technology had triumphed again.

Selenography, the study of the Moon, has gone through phases almost 400 years since the invention of the telescope. Through time, after a period of intensive professional work usually brought on by a technological change, interest in the Moon virtually disappeared from the professional ranks. The first technological advance was,

FIGURE 8-1   *A Surveyor 5 image, taken of the footpad; this craft landed in Mare Tranquillitatis. (NASA)*

of course, the invention of the telescope itself. After the pioneering work of Galileo, the first lunar map by Johannes Hevelius, and the work of other astronomers of the seventeenth century, the study of the Moon fell out of favor with professional astronomers. With the introduction of good achromatic refractors and micrometers for detailed measurement, the study of the Moon took off again in the early 1800s. Then, once again, the study of the Moon fell on hard times.

In the last decade of the nineteenth century, large telescopes combined with the new science of photography to signal the start of a new age of lunar exploration—and, unfortunately, an eventual decline in professional interest. Finally, the advent of space travel kindled the hopes of professionals who sought the ultimate platform to study the Moon: the Moon itself. That hope was realized six times between 1969 and 1973. We haven't gone back since; and even with smaller craft such as the Lunar Prospector and Clementine, darkness has fallen once again on the study of the Moon.

During these periods of inactivity, the quality of work done by amateurs rivals and sometimes surpasses that done by the professionals. During these lulls the Moon is returned to those who appreciate her most: amateur astronomers.

THE DOUBLE-PLANETARY SYSTEM

The Earth–Moon system is unique in the solar system and possibly in the entire universe. From Venus or Mars, Earth would appear to be a double star, a blue-green star with a smaller white attendant circling it. It has been postulated that if the Moon circled Venus instead of Earth, we would have a totally different type of civilization. Seeing the motion of the Moon around Venus would have been a great clue to the scientists of ancient Greece that the Earth revolves around the Sun. If that theory had taken firm hold 2,000 years before it actually did, who knows what would have happened. (But it didn't—and we celebrated an Earth-, or geo-, centric model of the solar system.) It has also been postulated if the Moon did orbit Venus, there would be no humanity on Earth to see it. Without the Moon, there would be no great tidal movements and thus no tidal pools from which life began its trek to land. Maybe it all worked out for the best.

For years, scientists believed we always saw the same face of the Moon because the satellite did not rotate. As with many things, these scientists were wrong. The Moon does rotate. Because the eastward rotation and the eastward movement of its orbital path around the Earth are equal, we just don't see the rotation. That is why it looks as if the same face is always turned toward Earth. But if you look closely, you may detect some rotational motion. You won't see anything over the course of a couple of hours, however, because the motion is very subtle.

If you follow the Moon carefully through more than one *lunation*—the complete cycle of its phases, which lasts just over 29.5 days—you will see this effect, which is called *libration*. Libration allows us to "see around the corner" of the Moon (Fig. 8-2).

The Moon has a very stable rotation rate. Because of this, its axis is stable with respect to the stars. On certain days during a lunation, the Moon's axis is tipped toward Earth by 6.5 percent. When that happens, we can see 6.5 percent past the pole on the other side of the Moon. These events are separated by 2 weeks, so we see the area beyond the Moon's north pole first and 2 weeks later we peek past the south pole. This is called *libration in latitude*.

The Moon does not follow a circular orbit around the Earth, but one that is slightly elliptical. Because of this, the Moon revolves slightly faster than it rotates

Figure 8-2 *Lunar libration. (Courtesy Lick Observatory)*

when it is closest to the Earth, or at *perigee*. When this happens we can glance around the eastern edge of the Moon and see about 8 percent more of its surface. On the other side of its orbit, when it is at *apogee*, or the point farthest from the earth, the reverse is true. Here the Moon revolves slower than it rotates and we get a look past its western edge, or *limb*. This effect is called *libration in longitude*.

We see more of the Moon than one would expect if only one side were turned toward us perpetually. You would expect to see 50 percent of the disk under those conditions. Because of libration however, we are able to see 59 percent of the disk. You can find out the amount of these librations by consulting certain publications, including the *Astronomical Almanac*, published yearly by the U.S. Naval Observatory, and the *Observer's Handbook*, published annually by the Royal Astronomical Society of Canada (RASC).

LUNAR GEOGRAPHY

When you look at the Moon with the naked eye, the surface types are very apparent. They form what has been called the "Man in the Moon," the "Old Woman," or even the "Rabbit" and "Frog" (Fig. 8-3). The dark areas you see are called *maria* (Fig. 8-4),

FIGURE 8-3 *The full Moon taken by the Clementine spacecraft on March 15, 1994 (the spots are from the startracker instrument). (NASA)*

FIGURE 8-4 *Mare Serenitatis is one of the many lunar maria. (Courtesy Lick Observatory)*

the plural of the Latin word for "sea"; *mare* is the singular. These are large flat plains formed by gigantic lava flows. If you look carefully, you will see that they are not black but a grayish color, something that becomes more evident with a pair of binoculars or a telescope. Contrasting with these dark areas are the lunar *highlands* (Fig. 8-5), mountainous regions containing countless craters. Everything else we see on the Moon is viewed against the backdrop of either a mare or a highland.

If the Moon is known for any one feature, it is for hundreds of *craters*. Ask people on the street what they think of when you say "Moon" and the odds are they will say "craters." The term *crater* is really a generic term. There are many subdivisions referring to crater type and size. *Impact craters* are caused by the collision of projectiles with the lunar surface. They have certain unique characteristics that make them easy to identify. They are usually deep and almost perfectly circular depressions, with steep interior walls and gently sloped exterior walls. There is also usually a splash effect from the impact—a blanket of debris called *ejecta* surrounding the crater. Impact craters usually have a *central peak* thrown up by the impact. Several excellent examples of impact craters are Copernicus and Theophilus. Another interesting impact crater feature is the *ray system*. The most prominent sys-

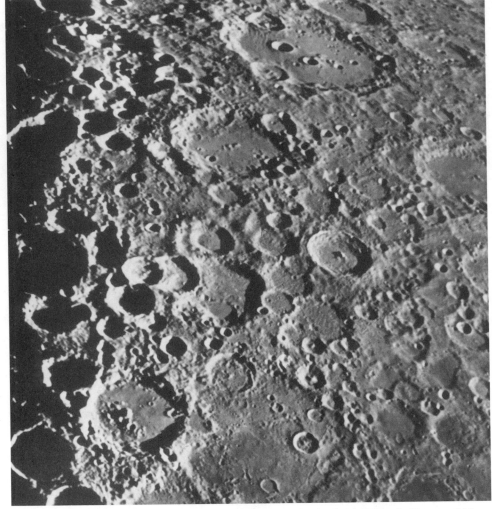

FIGURE 8-5 *The area surrounding the crater Tycho is an example of the lunar highlands. (Courtesy Lick Observatory)*

tem of rays, stretching 1,500 miles across the lunar surface (Fig. 8-6), belongs to the crater Tycho.

Volcanic craters are caused by, of course, volcanic action. They are characterized by flat, dark floors, similar to the maria. Both the interior and exterior walls are uniform, and there is no sign of ejected matter. The best example of a volcanic crater is Plato. There are also *walled plains, ring plains craters, craterlets,* and *crater pits* (in order of descending size).

In addition to craters, there are a number of other interesting features on the Moon. Primary among these are the *mountain ranges* that cross the surface. The finest example of a lunar mountain range is the Apennines, which stretch in an arc from just west of Plato, southeast along the rim of Mare Imbrium, to Eratosthenes. Near the

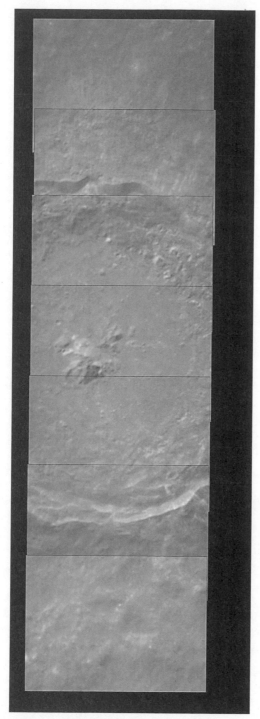

FIGURE 8-6   *The center of the crater Tycho taken by*
the Clementine spacecraft. (NASA)

Apennines are the Alps, another excellent example of a lunar mountain chain. Starting next to Plato, they stretch southwest to the crater Cassini. The Alps are broken up by another lunar feature, the *valley*: The Alpine Valley runs for 110 miles through the rough terrain of the Alps. Its floor is flat, and the valley is perfectly straight.

The Moon is also home to a number of smaller interesting features. *Rilles* appear as cracks in the surface. Most of them are small, but a few, such as the Ariadaeus rill near Mare Tranquillitatis, can run for hundreds of miles. While rilles are depressions in the lunar surface, *ridges* are upswellings. They are found near the edges of the maria and suggest waves of lava frozen in time. *Scarps* fall somewhere between ridges and mountain ranges. They are higher than the ridges, but not as high as the mountain ranges. Finally, we have small swellings in the lunar surface called *domes*.

These features of lunar topography might seen unique, but remember that every type of lunar feature has an analogy here on Earth. The Earth's features are subject to weathering whereas those on the Moon sit unchanged for millions and even billions of years. The famous Meteor Crater just outside Winslow, Arizona, is comparable to the lunar impact craters because it formed from the impact of an asteroid. Though it is eroded, you can plainly see the characteristic slope of the interior and exterior walls, and surrounding the crater is debris ejected when the object struck the surface. If you want to study lunar volcanic-like craters you can take a trip to Hawaii. (January is the best time to do this, especially if you live in the snow belt.) Look at Mauna Loa and the other craters in the island chain. While there, look for shield volcanoes as well; these are analogous to the lunar domes.

While these impact craters and volcanoes were formed by similar events on both the Earth and the Moon, mountains, mountain ranges, valleys, rilles, ridges, and scarps all have more site-specific origins. On Earth, such features were formed by events unique to the planet such as plate tectonics (the continual movement of the continental plates), and water and wind erosion. Similar features on the Moon were not caused by these events, but by both volcanic action and the impact of large objects with the surface.

TAKE THE TERMINATOR TOUR

As the Moon moves between the Earth and Sun at the beginning of each lunation, it catches the sunlight, which illuminates its surface. This illumination is the cause of the phases of the Moon we see from Earth. As the Sun rises throughout the lunar day, the *terminator*, or sunrise line, divides the area of the lunar surface illuminated by the Sun from the area still in darkness.

An excellent project for the beginning amateur is to follow the course of the terminator as it crosses the Moon. A project like this gives the beginner a chance to become more familiar with lunar features. It does not require a powerful instrument and can be easily done with a good pair of binoculars or a small telescope. That is one of the advantages to studying the Moon. You don't need a large instrument to enjoy the surface of our neighbor in space. In fact, many observers feel that the Moon is best seen with a small instrument so that it just fills the field of view. So let's take a trip across 238,000 miles of space and see what the Moon has to offer.

The detection of the earliest crescent Moon has become a challenge to amateur and professional observers alike. The only time that observers from Earth can actually see the Moon pass from the old Moon to the new Moon is during a solar eclipse when the Moon passes across the face of the Sun. The new Moon rises about sunset, and under certain circumstances, a thin crescent may be visible low in the western sky after sunset. Because of the size of this lunar crescent, its proximity to the glare of the Sun, and the fact that it sets so soon after the Sun, it is rare to see a new Moon earlier than 22 hours old.

The record for visually sighting this young Moon belongs to the famous lunar mapper and director of the Athens Observatory, Julius Schmidt. On September 14, 1871, Schmidt sighted the Moon when it was only 15.4 hours old. The record for seeing a young Moon belongs to an amateur from California. On March 15, 1972, Robert Moran observed the very thin crescent Moon when it was only 14.9 hours old. He used 10 × 50 binoculars to make his observation.

On July 14, 1988, the most recent opportunity to observe a young lunar crescent, Robert Stalzer, president of the Chicago Astronomical Society, journeyed to the observation deck of the Sears Tower in Chicago. From his vantage point, he had an excellent view of an unobstructed western horizon. There were clouds at sunset, but Bob began his search through a hole that pierced the cloud cover. Using 10 × 20 binoculars, Bob was able to detect what he estimated to be a 20-hour-old crescent. If not for the clouds, he felt sure that the Moon would have been visible to the naked eye.

The best time to make an attempt at seeing such a young Moon occurs under the following conditions: First, the Moon should be directly above the setting Sun. This occurs most often during the early spring and late winter. Second, the Moon should be at perigee. This is because at this point in its orbit, the Moon is moving rapidly away from the Sun—so the distance between the Moon and the setting Sun is greater than normal.

To see this youngest of lunar crescents, you need to have a number of things going for you. An unobstructed view of the horizon is essential. The target you are looking for is going to drop below that horizon in a matter of minutes as it follows the sunset. Any viewing time you can gain by not having trees or buildings in the way is to your advantage. You also need a sky with great transparency and darkness, because the thin crescent is a very dim object. Finally, you need a new Moon well separated from the Sun at the time you will be making your observation. You can find this out by plotting the positions for the Sun and Moon on a star chart and simply measuring the separation. You can check the *Observer's Handbook* or the *Astronomical Almanac* to see if conditions are ripe to catch a thin lunar crescent.

Now that you have bagged a very young Moon, it's time to see the show the Moon puts on over the course of a lunation. As the new Moon races eastward away from the Sun, its features become swamped with light from the "rising" Sun (from the perspective of the Moon) along the terminator. It is along this line that you will see most lunar features at their best. The sharp shadows thrown by rocky structures give them an almost three-dimensional effect. Whether you use a pair of binoculars or a telescope, the Moon will never let you down as a target for observation (Fig. 8-7 and 8-8).

We start our first tour—which will take in many nights of observing—with the 2-day-old Moon. There are a multitude of features for the observer to study. The most obvious feature is the dark oval shape that lies along the eastern limb of the Moon, Mare Crisium. Isolated from other lunar maria, Crisium is some 270 miles across from north to south and 350 miles across from east to west. It is an excellent

Figure 8-7 *The first-quarter Moon. (Courtesy Lick Observatory)*

FIGURE 8-8   *The last-quarter Moon. (Courtesy Lick Observatory)*

gauge of the libration effects—it seems to turn a fuller, rounder shape toward Earth during periods of libration in longitude.

The Moon has always been considered a dead world, unchanged over the billions of years since its formation. But something seems to take place on the lunar surface from time to time. These events are called *lunar transient phenomena* (LTPs). Since 557 A.D., over 1,500 reports of LTPs have been made. They fall into three distinct categories: First, there are periods when a feature or area is hidden from view. Second, color changes covering both large and small areas of the lunar surface are another type of LTP. Finally, there are glows and flashes of white or colored light. One of the regions in which these LTPs occur is Mare Crisium. Observers of the eighteenth century said that the grayish hue of the mare seemed tinted with green during the full Moon. Rev. T. W. Webb, a proponent of astronomy and author of *Celestial Objects for the Common Telescope*, reports that he and other observers saw the floor of Mare Crisium "speckled with minute dots and streaks of lights."

South of Mare Crisium, just below the lunar equator, is the crater Langrenus, surrounded by crater walls that rise 16,000 feet above the lunar surface. It is 85 miles in diameter and has a central mountain rising some 6,000 feet above the crater floor. This mountain can catch the rising Sun when the floor is still deep in shadow, causing it to gleam like a diamond against a black cloth.

Farther south you will find the crater Petavius. Petavius is a huge crater that is 110 miles across. With a small telescope you can easily see the terracing of the inner walls. The floor of the crater seems to come alive with the angled rays of the Sun. Three rille systems criss-cross the crater's slightly humped floor. Binoculars should show one system, but the others will require a telescope. The center of the crater is home to a mountain complex, the highest of which rises over 8,000 feet above the crater floor. The floor itself is very dark and devoid of craters.

Still farther south is the crater Furnerius. This large crater is 81 miles across and over 14,000 feet deep. Furnerius belongs to the class of craters known as *mountain-walled plains* or *walled plains*. Craters of this class are very large but relatively shallow. Because of this, it would seem that this crater should be related to Petavius, but they differ in significant ways. Petavius has a prominent group of mountains at its center; Furnerius does not. This is the principal way in which scientists define a mountain-walled plain.

Another characteristic of the walled plains is their floors. The floor of Furnerius is smooth compared with the rille system that crosses Petavius's floor. Some scientists speculate that at one time the entire floor of what is now Furnerius completely collapsed onto the then-hot magma below. Another piece of evidence supporting this theory is that Furnerius is the center of a bright ray system—as if the hot magma splashed up around the edges of the collapsed section and splattered across the landscape. Regardless of what caused the smooth floor and the ray system, it is evident that this is an old crater. Its walls are pocked with other craters, which shows that Furnerius is older than the intruding, fresher craters.

Moving inward from Furnerius you will come to the crater Rheita and the Rheita valley. Of the two features, the valley seems to be older. If you look carefully, you will see that the crater seems to overflow through the walls of the valley, an indication that the valley came first and the crater followed. The crater Rheita is smaller than the other features we have looked at up until now. It is only 44 miles across and is 14,500 feet deep. The valley runs for about 300 miles and is about 2,000 feet deep.

As the terminator moves west across the Moon, you will notice that the other maria, which are becoming visible, are connected. Each seems to flow into the next. One reason the old mappers thought them to be seas was because of this flowing connection, thus their name. Northwest from Rheita and the Rheita valley is a dark mare called Nectaris, or the Sea of Nectar. At the south end of this mare is our next stop, the crater Fracastorius. This crater is 73 miles across, but you will have to look closely to find it—the crater has lost most of its north wall. The remaining walls rise only 5,800 feet above the surface. Because of this relatively small height, Fracastorius doesn't throw too much shadow. In fact, this very old crater looks more like a bay than a crater.

Two other features frame Mare Nectaris for us on our guided tour across the Moon. The first of these is found along the eastern edge of the mare. The Pyrenees mountains fit better into the category of scarp than mountains. They rise above the "shore" of the mare and run for about 200 miles along its rim. Whether you want to classify this feature as a true mountain range or a scarp is up to you. Whatever you end up calling it, the Pyrenees are easy to identify and a striking object in the rising shadows thrown by the Sun.

Across the mare in its northwest corner, we come to our other framing feature, the crater Theophilus (Fig. 8-9). At 65 miles in diameter and over 22,000 feet deep, Theophilus makes an impressive sentry guarding the connection between the mares Nectaris and Tranquillitatis. Theophilus is the deepest of all lunar craters and has a central mountain peak reaching 7,500 feet above the crater floor. When near the terminator, the floor of the crater is lost in darkness while the tip of its mountain shines brightly.

Moving north and west from Theophilus we come to a small, seemingly insignificant crater called Arago sitting in Mare Tranquillitatis. The crater is the smallest we will see in this short trip across the Moon. It is only 16 miles across and 5,900 feet deep, but we didn't come here just to see the crater. Arago sits on the relatively flat surface of Mare Tranquillitatis, made famous by Neil Armstrong and company. If you look closely at the surrounding mare plain with a telescope, you may see little bumps every now and then. These are lunar domes—and the area is rich with them. Directly north of Arago, about the distance of the crater's diameter, is another dome called Arago Alpha; still another dome is directly west of the crater at about the same distance, called Arago Beta. (Lunar mappers label any surface bump with the name of its associated crater and a Greek letter.) Just southeast of Arago sits the old, rayed crater Lamont. Between Lamont and Arago are six objects suspected of being domes.

North of Arago on the shore of Mare Serenitatis is our next stop, the crater Posidonius. At 62 miles in diameter it has all the characteristics of walled plains like Furnerius, but with a distinct difference. Furnerius and Posidonius both have the flat floor that is one of the trademarks of the walled plains, but through a telescope you can see that Posidonius is a jumble. The crater floor is filled with numerous smaller craters, an indication of great age. The floor of the crater also contains four rille systems, one of which, Rima Posidonius II, is easy to spot with a small telescope.

Moving west around Mare Serenitatis you come to your first real lunar mountain range: the Caucasus mountains. These are immediately distinguishable from the

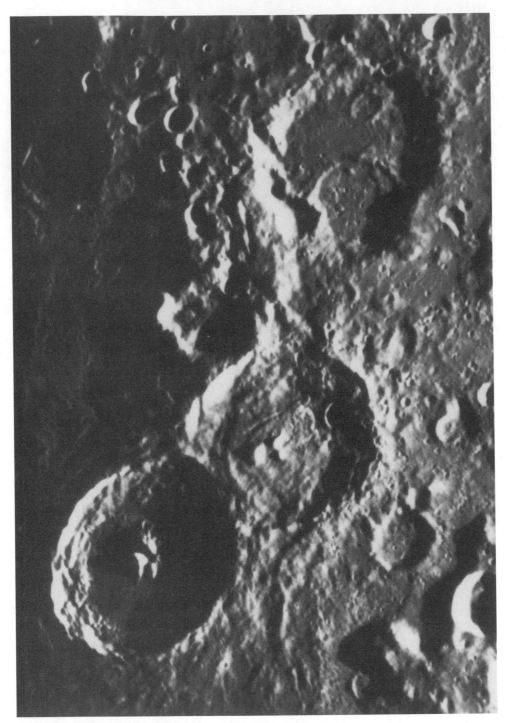

FIGURE 8-9 *The crater Theophilus. (Courtesy Lick Observatory)*

Pyrenees because they are almost twice as high. It is no wonder that some observers consider the Pyrenees to be nothing more than a glorified ridge. Before the terminator even reaches the Caucasus range you can see it. The rising Sun catches its peaks, causing them to shine against the still-shadowed Moon. This range runs for over 300 miles, separating Mare Serenitatis from Mare Imbrium.

The first half of the tour is coming to an end. The Moon is approaching first-quarter phase, and we have given 13 lunar features a detailed look. The last stop on this part of the tour takes us back toward the south end of the Moon: to Maurolycus, a crater 73 miles across and almost 17,000 feet deep. Its size and depth seem to dominate this area of the lunar surface, and when on the terminator, it stands in bold relief. It is a relatively young crater, plainly formed by impact. Observers of the nineteenth century reported that the floor of the crater was criss-crossed by up to a dozen bright lines when seen under high illumination. The central mountains of the crater are easily seen with binoculars, especially when it is near the terminator.

By now you have watched the terminator creep across the Moon at a rate of about 12 degrees per day for about 5 days. It is seven days past new Moon and our satellite has traveled through one quarter of this lunation, but there is much more to see.

We will begin the second half of our journey just south of the lunar equator at a group of three craters. The largest of the group is the crater Ptolemaeus, with a diameter of 93 miles and a depth of 9,800 feet. Looking at the crater's interior with a pair of binoculars shows that the crater floor looks smooth. Just north and east of the center of the crater, however, you will notice a small, bright spot that is the small crater Ptolemaeus A. Only 6 miles across, it is the smallest crater on our tour. Through the telescope, the floor of Ptolemaeus reveals a complex structure. Craters and craterlets are everywhere. The oblique light of the terminator reveals a wealth of ridges and depressions on the crater floor.

At the other end of the group is the smallest of these three craters, Arzachel. It is the youngest of the group and shows all the characteristics of a classic impact crater. It is 59 miles across and 13,000 feet deep, and has a central peak rising 4,600 feet. A look at the crater with a telescope reveals beautifully terraced walls over 10 miles thick.

The crater in the middle of the group, however, is the most interesting. Alphonsus is an old crater that has been partially flooded with lava, possibly thrown up by the impact of the object that created Arzachel. It is 73 miles across and only 6,600 feet deep with a central peak only 3,800 feet high.

But it is not its physical characteristics that make Alphonsus an interesting object. In 1956, the lunar expert Dinsmore Alter used the 60-inch reflector at Mount Wilson to make a photographic study of the Alphonsus region. The photo of the area east of the central mountain was puzzling because it showed the floor of the crater covered by a patch of haze. A Russian astronomer, Nikolai Kozyrev, became interested in the problem of lunar changes and began observing the region with the Crimean Observatory's 50-inch reflector in November 1958. Some three months later, in *Sky & Telescope*, Kozyrev revealed that he did see a change take place—and he had definite proof in the form of spectrographs. Another LTP was found.

While at the telescope and photographing the reflected the spectrum of light from the central mountain, Kozyrev saw what looked like a gas cloud form near the central peak, accompanied by a reddish glow. He observed a second event during which the peak of the mountain became very bright and white. A later analysis of the spectrographs showed the presence of a gas cloud made up of carbon molecules.

During the mid-1960s, Alphonsus was almost visited by Ranger 7 (Fig. 8-10); it then became the target of the last Ranger probe, Ranger 9, which impacted the crater on March 24, 1965. Photos returned by the probe showed certain parts of the floor covered by a dark material that could be ashes and cinders thrown up by a lunar volcano. Could something like this have caused the events that Kozyrev observed?

The next feature on our tour is just southwest of Arzachel. Many people, when they first see this feature, are struck by its artificial look. The Straight Wall is just what its name implies: a straight wall of rock that runs for 70 miles, rising from its north end to a height of 1,200 feet, maintaining that height for 20 miles, and then dropping back to the plain. The sharp figure cut by the Straight Wall was caused when the plain on the western side of the feature slid away, probably because of volcanic action. The Straight Wall is thus the demarcation line between the two sections of the plain.

FIGURE 8-10 *This is the first image taken by the U.S. spacecraft Ranger 7 in July, 1964, 17 minutes before it impacted on the lunar surface. The larger crater at center right is Alphonsus. (NASA)*

Because the western side of the Straight Wall is lower than the eastern side, the feature presents a changing image to us during the long lunar day. During the period of the morning, or rising Sun, the Straight Wall will appear as a dark streak on the lunar surface because the shadows are thrown from the ridge of the wall onto the lower western plain. During the period of the setting Sun, or evening, the light illuminates the exposed face of the wall and it appears as a bright streak.

If you extend the line of the Straight Wall to the northwest, you will come to our next stop and what many think is the most spectacular crater on the Moon: Copernicus (Fig. 8-11). This crater stands out even when the Moon is viewed with the naked eye. With a little optical aid from binoculars, the crater begins to dominate the lunar landscape. With a telescope it is simply magnificent.

FIGURE 8-11 *The craters Copernicus and Eratosthenes. (Courtesy Lick Observatory)*

Copernicus is, by all measures, a rather mediocre crater, with a diameter of 60 miles and a depth of 12,600 feet. You have already seen craters that are bigger and deeper. What makes Copernicus so special is that it's a young crater. Compared to a structure like Fracastorius, Copernicus is the new kid on the block. The relative youth of the crater gives it a youthful appearance: Its lines are strong, sharp, and clean, not like the collapsed wall on Fracastorius. As an excellent example of an impact crater, Copernicus has a fine ray system that was thrown up when its meteoric creator collided with the Moon. Combined with its high surface brightness and its position on a dark plain, the ray system seems to draw attention to the crater. Finally, Copernicus obeys the three basic rules of real estate: location, location, location. Take a look at how Copernicus is placed on the lunar disk. Not very far from the center, is it? Even if you aren't looking for it, your eye is going to automatically gravitate to that bright patch near the center of the Moon. It's almost a reflex action.

Close by Copernicus, just to the northeast, is the crater Eratosthenes. Though smaller, at only 37 miles across and 12,000 feet deep, Eratosthenes is similar to Copernicus in many ways. Because the Sun strikes Eratosthenes before Copernicus, it is sometimes confused for the larger crater. The area between the two craters is interesting to study with a telescope. Because both of these features are impact craters, a great deal of material was ejected during impact. The dozens of tiny craterlets that fill the plain between the two craters tell the story of where that material came down. Eratosthenes also has an interesting feature to its southwest—a 50-mile-long ridge that looks like a small tail.

Just east of Eratosthenes and stretching northeast along the rim of Mare Imbrium is the finest example of a mountain range on the Moon. The Apennines (Fig. 8-12) extend for 450 miles from Eratosthenes to the edge of Mare Serenitatis and rise to a height of 20,000 feet. The peaks catch the Sun well before the arrival of the terminator and hold those rays after the terminator has carried sunset away from the region. The mountains are easily seen under the full glare of the Sun during a full Moon.

The eastern approaches to the great flat, dark plain called Mare Imbrium are guarded by two great sentinels. To the south is the large, dark-floored crater Archimedes. Another walled plain, Archimedes is 51 miles across and 6,800 feet deep. Its walls are low at the north and south points of the ring, almost as if some projectile went through them on a straight line. The floor of the crater is dark; the selenographers William Beer and John Madler called it "smooth as glass" in their epic 1837 work *Der Mond*.

At the north end of the approach to Mare Imbrium lies another great walled plain, Plato. Although many observers consider Copernicus to be the premier lunar crater, Plato often ranks as a close second favorite. Elliptical in shape, the crater is 64 by 67 miles across and 8,000 feet deep. The floor is very dark in contrast to the floor of Archimedes. Most of these old walled plains tend to fade out in the glare of the Sun as the Moon becomes full, but the floor of Plato seems to get darker as the Sun climbs in the sky. Consequently, it is as easy to see at lunar noon as it is at sunrise.

The wall structure ringing Plato is old and not as high as that of Copernicus. But as the Sun climbs over the wall during sunrise, it casts wonderful dark shadows across Plato's crater floor. There are sharp points and valleys that never cease to amaze the viewer. Plato has also been the sight of numerous LTPs, ranging from flashes and lights to obscuring fogs—so keep an eye out.

FIGURE 8-12 *The Apennines and the crater Archimedes. (Courtesy Lick Observatory)*

Between Plato and Archimedes lie two of the Moon's rare single mountains. Piton is directly south of Plato, and Pico is a bit east of a line joining the two craters. Piton covers an area of almost 190 square feet and rises 8,200 feet above the floor of Mare Imbrium. Pico is a more complex structure over 18 miles across at its base and jutting upward approximately 8,000 feet above the mare.

Back across Mare Imbrium and southwest from Copernicus you will find the crater Gassendi. Seventy miles across and 6,600 feet deep, Gassendi offers a great deal of detail especially for someone with a telescope. Lodged in its northern wall is a much smaller crater, Gassendi A. The floor of the crater is covered with rilles and a central peak that rises 3,600 feet.

Finally, the terminator reaches the crater Grimaldi just before the Moon becomes full. Grimaldi is a very large crater over 140 miles across and more than 10,000 feet deep. Remember how dark Plato's floor was? Grimaldi's is even darker. It only reflects back 6 percent of the light that reaches it.

The Moon has reached the phase we call *full*, and for a brief period there is no terminator. But we're only half done. We saw sunrise across the face of the Moon. The features we saw at the start of the tour—Langrenus, the Rheita valley, and Theophilus—were touched by the rays of the Sun slanting in from the east. In effect, we have only seen half the show. Now it's time to watch as the Sun drops toward the western limb of the Moon and the shadows stretch out from that side of the craters. It's going to be a different show.

In one sense, the sunset tour is going to be easier because you already know the craters and other features to watch, but in another sense it's going to be harder. Because of the way the Moon orbits the Earth, it rises about 45 minutes later each night during a lunation. When you began the tour, the Moon was up just after sunset, but it has been falling behind in the sky later each night. Now that the Moon is past full, it will be rising later each evening, so it's going to take a little discipline to continue the tour. By the time this lunation is over, you will be observing in the wee hours of the morning before sunrise. But it's worth it. The shadows you saw thrown across the floor of Plato were cast by the east wall of the crater.

If you're just a tiny bit curious about what kind of shadows its west wall throws, it's time to watch the moon in reverse. Take the time to go "backward" through the text, observing and noting all the differences you see as the Sun's light crosses the Moon in the opposite way. The differences are there, caused by the different angle of light on the same craters, highlands, and rills.

THE MOON AND THE SERIOUS OBSERVER

So now you have a feel for some lunar geography. You took the entire terminator tour and can find your way around the lunar landscape pretty well. But there is more to observing the Moon.

MAP AND FILTERS

If you decide that you want to do some serious lunar observing, there are a couple of things that you will need. The first requirement is a good lunar map. You have an idea of what the various lunar features look like, but you need a means of finding your way through the hundreds of thousands of craters that dot the Moon.

Sky Publishing Corporation distributes an excellent resource titled *Lunar Quadrant Maps*, prepared by scientists from the University of Arizona's Lunar and Planetary Laboratory. Printed in four sections, the maps show all the named features on the visible face of the Moon down to a size of two miles. Each chart is large, measuring 23 by 27 inches. The set is rolled and mailed in a heavy tube, so it is a good idea to mount them on some sort of board for use at the telescope. You can even divide each chart into six sections, photocopy each of those sections, put the copies into transparent sleeves, and then put them all into a ring binder. Then mount the original charts on a wall or bulletin board, readily available for when you go to the telescope—or for times you want to "study the Moon" in the daylight.

You night have noticed that looking at the Moon, especially when it is near full, is very tiring on the eyes. This is because of the tremendous glare reflected from the lunar surface. The easiest way to overcome this is to use a filter.

A number of companies manufacture special filters for observing the Moon. They are threaded at both ends and easily screw onto the end of your eyepiece. The best filter for reducing glare is called a *neutral density filter*, a gray filter that has no effect on the image other than to reduce the overall glare. Also available are *variable density polarizing filters* that change their darkness to admit more or less light to suit the observer.

> *Note! Neutral density filters or polarizing filters may look dark, but they are not designed to look at the Sun. Do not use neutral density filters or polarizing filters to view the Sun!*

A good source for these filters is a photo shop. Eastman Kodak and many small companies manufacture lines of filters for photographic use. They are much bigger than the filters made specifically for an eyepiece, but that's all right. It's just as easy to hold the filter between the eyepiece and your eye as it is to have a specific eyepiece—and it can be cheaper, especially if you get a used photo filter. (*Warning: We are not talking about a filter for the Sun here, only for planets and the Moon*).

Wratten filters are simply sheets of filter material that you can cut to a convenient size and place between your eye and the telescope's eyepiece. If you are studying LTPs, you can use a red (Wratten 25) and a blue (Wratten 38A) to bring out the contrast in the suspected LTP. Wratten filters are expensive because they are usually sold in large, 3-inch-square sheets. Contact a local photographer and ask if he has any in his dust bin. He probably does and would be happy to get rid of the torn and broken things. Take the filters or the pieces home and mount a red and blue piece side by side in a 2 × 2 slide mount. You can place the slide mount up to your eye and move it back and forth, which will cause any suspected LTP to flicker.

STUDYING LUNAR DOMES

Now that you have a good lunar map and a filter to cut the lunar glare, what can you do? A project you can start even before you have a telescope is a study of lunar domes. Lunar domes are those swellings we saw around the crater Arago in Mare Tranquillitatis. They are important to lunar science because they may represent a late stage of volcanic vents that could have contributed to the formation and later changes of the lunar maria. One of the by-products of the technology that took man to the Moon in 1969 was the complete photographic portrait made by a satellite preceding the landing. The Lunar Orbiter project of the late 1960s still offers a wealth of detail that anyone can study.

The *ALPO Lunar Dome Recorder* can supply you with a listing of suspected lunar dome sights plotted on the *Lunar Quadrant Map* discussed earlier. An observer can then study such atlases as *The Moon as Viewed by Lunar Orbiter* (NASA Publication SP-20) and the *Atlas and Gazetteer of the Nearside of the Moon* (NASA Publication SP-206). Although these works are out of print, they can be found at some libraries. There are also a number of atlases created from Earth-based photographs of the Moon, but these are not necessary. And the best part of observing the Moon with these books is you don't have to worry about being clouded out.

The procedure for the project is simple and can be rather enjoyable. For equipment you will need a magnifying glass and a ruler equivalent to the scale of the photos you will be studying. Begin by locating the area on the ALPO publication that you want to chart. Each suspected dome is listed with a number, so you don't have to worry about figuring out any sort of coordinates for any objects you find. Flip to the section of the atlas that has the area you wish to search and carefully go over that area with your magnifier. Take your landmarks from the ALPO chart. If you find an object that corresponds to a suspected dome on the ALPO chart, note its number and look for the following details:

- Is the suspected dome in an area of lunar highlands, maria, or somewhere in between?
- What is the object's approximate size? Use your ruler for this. Is it less than 5 kilometers (km) at its longest? From 5 to 20 km? 20 to 35 km? Or larger?
- What is the object's shape? Circular? Elliptical? Polygonal? Irregular? Or is it too ill-defined to classify?
- Does the object have a gentle or moderately sharp slope?
- What would the object look like if you were to view it in cross section? It is hemispherical? Would the summit be flat or sharp, or would there be multiple summits? Is the suspected cross section regular or off center?
- Are any details visible on the summit? Is it a craterlet or pit? A ridge or a peak? A cleft or valley? Or even nothing at all? How is any detail positioned on the summit? Centered? Off center? On the edge? Does the detail cross the entire summit?

Later, you can take your notes and charts out to your telescope and see if you can find the objects you found in the photos on the Moon itself.

MAKING A DETAILED STUDY

Once you obtain your telescope and become familiar with it, an enjoyable and worthwhile project for the beginning observer is a detailed survey of a specific area of the Moon over the course of an entire lunation. A project like this will give you valuable experience in planning and executing a scientific study that doesn't have to rely on expensive equipment. It will also help to sharpen your observational skills, something that every amateur astronomer can afford to do.

The area that you select for your study is up to you. Try not to select an area too packed with details, or you might have trouble keeping everything straight. For our example, the area of Mare Imbrium around Plato is interesting to study. Let's begin by selecting the exact area in which we are going to conduct our study by examining a high-resolution photograph of the region. One nice thing about being an amateur astronomer is that there is no pressure in anything you do. You can make this as ambitious or as simple as you want; it depends on you.

If you are going to study the crater over a period of a complete lunation, you must have some way of recording the results of each evening's observations. There are a number of ways to record your observations. Every amateur keeps a notebook or a journal with his or her astronomical observations. Some use preprinted forms, some use a simple 4 by 6 card for each observation, and some use a loose-leaf notebook; others record their information on the computer (you can even use a database or word processing program). No matter how you record the information, it is very important to keep track of what you see and do in observational astronomy. Through your records, you will be able to trace your development as an observer and study both your weak and strong points. The simplest way to record an observation, be it of an area on the Moon or a distant galaxy, is by putting pencil to paper and writing.

Notebooks everywhere are filled with notational sketches such as the following:

February 12: 02:45 UT (8:45 P.M. CST) Having a bit of trouble locating R Cnc tonight, my diagonal always confuses me. Got to remember which way is celestial north.

or

June 26: 04:00 UT (9:00 P.M. CDT) Plato looks great tonight, the floor is very dark. Piton seems unusually bright tonight.

Many times these verbal descriptions are supplemented by simple line drawings made at the telescope. These drawings don't have to be the work of a Rembrandt. All the amateur wants them to do is convey an image of what was seen.

The best way to record lunar detail takes one of two forms: a photograph or a drawing. At first glance it would seem that a photo of a feature is the way to go, but many observers don't feel that way. The human eye may lose out in a contest for seeing a faint point of light, but it is infinitely better at filtering fine details in a scene. A photo of a lunar crater is a single instant in time, whereas a drawing is a compilation of many fine details seen by the eye and transmitted from the brain to the paper.

"But I can't draw. I have trouble sharpening a pencil," you may say. You're not alone. So the easiest way to draw lunar features is to cheat a little bit and start with an outline chart of the feature you are going to observe (Fig. 8-13). Find a good-quality photo of the target area such as the one in the NASA atlases or a photo taken by an observatory. Decide which area of the photo you are going to observe, and carefully trace the basic outlines of the features you want to study. At the telescope, use pencils of varying sharpness to record the fine details you see.

You're not cheating. This technique was pioneered in the 1890s by a German selenographer named J. N. Krieger. He used this technique to compile one of the most accurate, and beautiful atlases of the Moon ever published: *Joh. Nep. Kriegers Mond-Atlas—Neue Folge*. Published in Vienna in 1912, the *Mond-Atlas* contains 58 plates of various sections of the Moon. The atlas is so rich in detail, it even rivals photographic atlases.

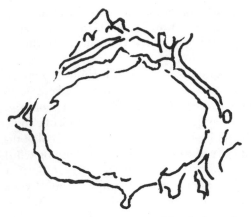

FIGURE 8-13 *Outline of the crater Plato for lunar drawing.*

Once you have your basic outline, make a dozen or more photocopies of it—one for each night that the feature is visible. Each time you observe your target, take along a blank outline and fill in the fine details as you observe. Somewhere on the form you should note the date and time, preferably in Universal Time, of the observation. Also record information about the telescope, eyepiece, and magnification used. Finally, estimate the seeing and transparency for the right. Use an area somewhere on the face of the outline chart to record notes about what you are seeing to supplement the information in the drawing.

Now that you have your outline chart, pencils, and telescope, you're ready to go. But what are you looking for? You can consider the basis of this exercise as being twofold. First and foremost you simply want to take some time and record what you see at the telescope. You want to become familiar with your target crater. Once you have found your target, take a few minutes and let your eyes roam around a bit. Remember what we said about talking to yourself? It's time for that again. How does the crater look in overall appearance? Can you make out anything on the crater floor? What kind of detail can you see in the walls? How about the surrounding rough terrain? Any small cracks or clefts? Don't just talk to yourself, make notes. Get in the habit of recording everything you see. Some observers take a small tape recorder out to the telescope—an easy way to record descriptions.

Another thing you can do while at the telescope is to monitor your target regions for LTPs. These can occur anywhere on the lunar surface, but many of the reported events seem to occur repeatedly in particular areas. ALPO recommends an extended study of eight areas for possible LTPs. They include Alphonsus, Aristarchus, Atlas, Copernicus, Herodotus, Tycho, Theophilus, and Plato. You will notice that our target for this exercise, Plato, is on ALPO's list—so it's not a bad idea to start including searches for these events in your observation of the area.

STUDYING OCCULTATIONS

Another area in which the amateur can make a significant contribution is in the study of *occultations*. An occultation occurs when the Moon (or another planetary body) moves between a star and an observer on Earth. Since the invention of the telescope, astronomers have used occultations to accurately determine the position of the Moon in its orbit, in respect to the time and location of the observer on Earth. Occultations also can reveal faint double stars and even help to determine the diameters of stars—a nice set of astronomical observations for an observer with a small telescope.

For every occultation there are two possible types of events. Depending on where the observer is located, he or she can see either a total or a graze occultation. A *total occultation* will occur if the observer sees the star disappear completely behind the bulk of the Moon. If the occultation is scheduled to take place along the northern or southern limb of the Moon, a *graze occultation* takes place. If the observer is positioned to watch the occultation along the Moon's limb, the star may disappear and reappear among the mountains and valleys along the lunar limb. The role of the observer during total occultation is simply to record the time of the star's entry and exit behind the lunar disk. During a graze, the observer should make a series of timings, one for each entry and exit the star makes as it moves along the uneven lunar limb.

The equipment needed for this work is simple. A telescope with a good mount is, as with any other type of astronomical observation, mandatory. It would be best to use an equatorial mount because this allows the observer to follow the target star more easily. A clock drive is not required, but it does allow you to focus all of your attention on timing the event and not worry about the star and the Moon drifting out of the field.

The next piece of equipment you will need is a radio that can be set to receive accurate time signals. The National Bureau of Standards (NBS) maintains a 24-hour radio station that does nothing but broadcast time signals. The call letters for this station are WWV, and it can be received on the shortwave bands of 5, 10, 15, and 20 MHz. If you use a tape recorder, you can make your comments and observations freely while the time signals from WWV are recorded in the background. There are also relatively inexpensive clocks now set up to receive signals directly from the NBS, broadcasting the exact time.

Finally, a stopwatch serves as a timing instrument and a backup. That's all the equipment it takes to get into occultation timings. The procedure for observing and timing a total occultation is just as simple. With the radio set to WWV and the tape recorder running, simply give a mark that can be heard over the time signals when the star either disappears or reappears.

It is easier for the beginning observer to watch and time total occultations. It is recommended that you try a few of these before you jump into a graze project. Predictions for these events for stars brighter than magnitude 5 appear in the January issue of *Sky & Telescope* and in the *Observer's Handbook*, published yearly by the Royal Canadian Astronomical Society (RCAS). The detailed prediction tables listed in the RCAS publication can tell you if the event is a disappearance or a reappearance against the bright or dark limb of the Moon.

The easiest occultations to observe are disappearances that take place against the dark limb of the Moon—which allows you to find the star well before its time of disappearance. An event that is to take place against the bright limb of the Moon is more difficult to observe because the brightness of the lunar limb makes it almost impossible to locate the target star—even if the observer begins to look for it well before the event.

Predictions for graze occultations can be obtained from the International Occultation Timing Association (IOTA) or the U.S. Naval Observatory; in addition, some astronomy magazines, including *Sky & Telescope* and *Astronomy*, will list major occultations on their pages. Observing and timing a graze is a bit more complicated and something best done by a group of people. A graze occultation can only be seen from a narrow path, often less than a mile wide. Because of this, it is necessary to place observers all along a predetermined length of the path. Many astronomy groups regularly observe these events, and it would be best to take your first plunge into graze occultations with a few lifeguards present.

9

OBSERVING THE PLANETS

The planets are seen very rotund, like little full moons and of a roundness bounded and without rays.

GALILEO

In July 1976, Viking 1 transmitted the first television pictures to Earth from the surface of Mars. Viking 2 followed two months later. These told astronomers, both amateur and professional, that the days of Earth-based telescopic studies of Mars were over.

In early March 1979, Voyager 1 reached the vicinity of Jupiter and returned fantastic pictures of Jupiter and its moons. In July 1979, Voyager 2 repeated the triumph of its twin spacecraft. Voyager 1 and Voyager 2 then hurried toward Saturn, arriving there in November 1980 and August 1981, respectively. Again, they told both amateur and professional astronomers that the days of Earth-based studies of Jupiter and Saturn with telescopes were over.

During the late nineteenth century, astronomy concentrated on the study of the planets. The moons of Mars and a new moon orbiting Jupiter were discovered by visual observation. Theories about the possibility of life on Mars abounded, and discoveries of comets and asteroids were an almost daily occurrence. The study of galaxies lay beyond the grasp of the instruments of the time, and theories of cosmology lacked observational data that made sense.

The development of the large reflectors and instruments such as the spectrograph and photometer at the beginning of the twentieth century changed the flow of astronomical thought. Interest shifted to objects millions of light-years away from the solar system. Theorizing about the formation of the universe took the place of wondering about the planets.

With the advent of the space age and the flights of robot probes to the residents of the solar system, it would seem that all the questions have been answered and nothing remains to investigate. Nothing could be further from the truth. In astronomy, as in any other branch of science, as old questions are answered new ones are raised. What is the reason for the massive dust storms that blanket Mars from time to time? What mechanism forms storms in the atmosphere of Jupiter? How are the orbits of Saturn's inner satellites and its ring system related? These are just some of the questions raised by the flights of the Vikings and Voyagers. They supplied some of the information necessary for the planetary answers, but Viking and Voyager were there for only a brief time. Not long after taking images of the red planet, the Viking project was terminated, and both Voyagers are now almost completely out of our solar system.

So where will the data to answer these questions come from? "Time on our large, complex telescopes is too valuable to waste on such trivial objects as the planets," say professional astronomers. They are concerned with analyzing the light from a distant quasar. "I guess we'll just have to wait until we can get another probe out there."

"But our budget has been cut to pieces," replies NASA, one of the only current ways to launch craft to the planets. "We may be launching smaller and cheaper craft to the planets, but in the case of the outer planets, it could take years to get a craft in orbit."

"Excuse me," interrupts the amateur astronomer. "Would data on the last ten Martian dust storms be more useful to you if I put it on a high-density computer disk?" This amateur is concerned about getting home in time to observe a new dust storm on Mars he heard about from his friend, an observer from Japan.

This may be a fictitious conversation, but its implications are true: Amateur astronomers watching the planets can fill in the blanks of planetary data.

Take Don Parker, the executive director of the Association of Lunar and Planetary Observers (ALPO)—and one of the best-known amateur astronomers, especially for his efforts to watch the planet Mars. Along with others in ALPO, he has been gathering data on the Martian Northern Polar Cap for years; ALPO members have been recording the shrinking and expanding of the cap since 1962. They have also produced some of the only weather data over the long term, taking down such items as weather trends and clouds. Even with craft such as the Mars Global Surveyor—which started mapping the red planet in the late 1990s—amateurs are needed. According to Parker, the MGS holds a polar orbit only. Thus, the rest of the planet was continually watched by amateurs to "fill in the data gaps."

THE AMATEUR'S SOLAR SYSTEM

Before the invention of the telescope, the solar system was composed of the Earth, Sun, Moon, and five planets (Mercury, Venus, Mars, Jupiter, and Saturn). By the time the telescope had been on the scene for some 240 years, astronomers had added two more planets, some planetary satellites, an occasional comet, and some asteroids to the Sun's family. In 1930, American astronomer Clyde Tombaugh added the planet Pluto to the stable. Ask any school child today about the solar system. You will hear a litany of the planets, with a couple of asteroids and Halley's comet thrown in for good measure.

A prominent astrophysicist once put it all in perspective. He said, "The solar system consists of the Sun, Jupiter, Saturn—and assorted rubble." From the standpoint of the astrophysicist, this is quite true: 99 percent of the solar system's mass is concentrated in its three largest bodies. The other planets—including Earth, the satellites of the planets, the asteroids, the comets, and everything else—compose less than 1 percent of the solar system.

It makes you think. Each period of history and each scientific specialty has its own view of what composes the solar system. And interestingly enough, each view is correct.

The solar system might consist of the nine major planets, but let's look at them from an amateur astronomer's observational point of view. We'll start with the inner planets.

Closest to the Sun, Mercury is a small world, only about 3,000 miles across. Called an *inferior planet* (Fig. 9-1) (a planet with an orbit inside the Earth's), it shows phases just like the Moon's when viewed through the telescope. Because of its orbit, Mercury is one of the most difficult of the planets to observe. When Mercury is east of the Sun, we see it in the evening sky after sunset. When it is west of the Sun, we see it in the morning sky just before sunrise. But because Mercury travels in such a tight orbit around the Sun, it never rises more than 30 degrees above the horizon and doesn't get much brighter than magnitude –1.5. Under the most favorable conditions, and because the planet is so low, you are looking through much of the Earth's atmosphere—where seeing is guaranteed to be bad. So it's always hard to find Mercury in the sky. Even when you do get to see it telescopically, it is never larger than about 10 arcseconds in diameter. It's a challenge to view, but most of us scratch Mercury from our planetary observing list.

Next out from the Sun is another inferior planet, Venus. Venus is almost the same size as Earth, so it is much larger than Mercury. At its farthest point from Earth, called *superior conjunction* for inferior planets, Venus is about 10 arcseconds across—as large as Mercury is when it is closest to Earth. At its closest point, called *inferior conjunction*, Venus is almost a full arcminute in diameter. Like Mercury and the Moon, Venus also shows phases, but its orbit brings it farther from the Sun and it rises to 47 degrees above the horizon. In addition, sunlight reflected from Venus's whitish-cloud atmosphere causes it to outshine everything in the sky except the Sun and Moon. Because it shines at between magnitudes –3.7 and –4.5, it is hard to miss this planet—and sometimes it can even cast shadows at night. Easy to find, Venus must be on the amateur's observing list.

Next we come to the Earth, a very difficult planet to view telescopically. Let's leave it off our list.

Outside the orbit of the Earth we come to the realm of the *superior planets* (Fig. 9-2) (those planets outside the Earth's orbit)—the first being Mars. About half the size of Earth, Mars moves around the Sun in an orbit that is much more elliptical than the Earth's. When it is opposite the Earth, at *conjunction*, Mars presents a very tiny disk that is less than 10 arcseconds across. When it is closest to us, in *opposition*, it can be as large as 25 arcseconds in diameter. Its size at opposition depends on its location in its elliptical orbit—thus it can be as small as 14 arcseconds or as large as 25. Even though Mars can be quite small, it is still easy to find with the naked eye. Its magnitude range is from about +2.0 to almost –3.0, and it has a reddish color. Mars is unique among the planets, as it is the only planet showing us a solid, undistorted surface. So again we have a planet that is easy to find and has something to offer. On the list it goes.

Past the orbit of Mars (and the asteroid belt) are the next planets for our list: the *outer* or *gaseous* planets. The first one is Jupiter, the largest planet in the solar system.

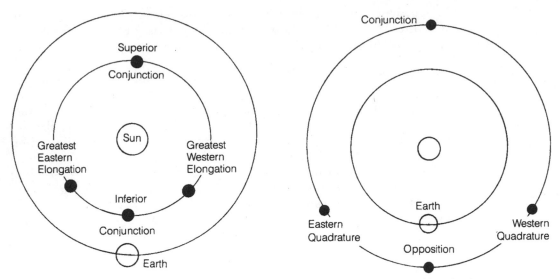

FIGURE 9-1 *(Left) The orbit of an inferior planet like Mercury or Venus. At inferior conjunction, the planet will appear as a thin crescent. At superior conjunction, it will appear as a full disk. At either elongation, it will appear half-full.*

FIGURE 9-2 *(Right) The orbit of a superior planet such as Mars, Jupiter, or Saturn. At western quadrature, the planet rises at midnight and sets at sunrise. At eastern quadrature, the planet is on the meridian at sunset. At conjunction, the planet is hidden behind the Sun.*

Easy to find, never below magnitude −2.0, and over 10 times the size of Earth, Jupiter always presents a grand telescopic sight with its cloud belts and satellites. This is a must-observe item for our list.

Finally, we come to the last planet known to science before Herschel added Uranus in 1781. Saturn is probably the most beautiful sight you will see in any telescope. Including its ring system, Saturn presents a target of between 30 and 40 arcseconds from tip to tip, the ball of the planet being slightly smaller. Why say anything more? Saturn is on our list, no questions asked.

Uranus is the first of the modern planets—those added after the invention of the telescope. Just at the naked-eye limit of magnitude 6, Uranus is a very large planet, measuring four times the size of the Earth. It is also very distant and never appears bigger than 4 arcseconds across. Although it is relatively easy to find, it never shows anything more than a green, featureless disk. Most of us cross Uranus off our list.

Neptune is another distant child of the Sun. Although it is large, at four times the size of the Earth, Neptune is so far away that it only presents a blue-green disk less than 3 arcseconds across. Again, not for us.

Finally we come to Pluto, the farthest outpost of the solar system. Pluto is so small and far away that it never gets brighter than magnitude 14. While you have at least a chance of glimpsing every other planet in the solar system, Pluto is beyond the reach of almost all amateur instruments. Even if you do find it, it will show no disk—thus it is definitely not for our observing list.

That is the solar system through the eyes of an amateur astronomer. It doesn't seem like much—just four planets: Venus, Mars, Jupiter, and Saturn—but you could spend a lifetime studying only one of them. Now let's see what an amateur can do with these planets.

CONSIDERATIONS FOR THE PLANETARY OBSERVER

A big factor in successfully observing the planets is your telescope. Each type of telescope has its own advantages and disadvantages concerning planetary observation. For years, the measure of a good planetary instrument was its focal ratio. Longer-focal-ratio telescopes, f/12 up to f/15 and sometimes even longer, give the instrument a small, dark field of view that creates greater contrast between the planet under observation and the background sky. This allows for more detail to be seen on the planetary surface. Combined with a medium to large aperture and as unobstructed a light path as possible, long focal ratios and large apertures usually allow for excellent detail resolution.

Many amateurs believe that the best instrument for planetary observation is the *refractor*. It nicely combines an unobstructed light path with long focal ratios. The problem with the venerable refractor is aperture. Most instruments over 4 inches in aperture are, to say the least, rather bulky. For instance, a 6-inch refractor with a focal ratio of f/15 is 90 inches long. It is rather difficult to transport, but not impossible. An 8-inch f/15, however, has a focal length of 120 inches—difficult to fit in the average car. The mounting required for instruments this size borders on the huge, and the larger the mount, the less steady—and thus less stable the image.

There are ways around this problem of length. The simplest solution is to bend, or fold, the light path by using high-quality optical "flats." Though some contrast is lost, a folded refractor with a 4- or 5-inch aperture and a focal ratio of f/15 or even f/20 becomes a manageable instrument—on a mount that doesn't take an Olympic weight lifter to move.

On the flip side of the coin are the shorter-focal-length refractors. For years, a refractor shorter than f/15 was considered unsuited for planetary work. The chromatic aberration given off by a lens of that focal ratio guaranteed an image with pretty colored rings around it. But technology does have its advantages. Computer-designed lens systems and special lens coatings now allow for the construction of shorter, effective focal ratios.

The primary problem with refractors is their cost. A good 6-inch lens in a cell and tube, but without a mount, can cost over $2,000. An 8- or 10-inch lens is out of sight for most amateurs.

Long focal lengths are also a problem with *reflectors*. The common newtonian design used today has a focal ratio of about f/5. According to many planetary specialists, this is just short of the optimum focal ratio for good contrast in a planetary image. A newtonian that will be used for planetary work should have a focal ratio of f/6 or longer. But in astronomy and optics, just like everything else, there are exceptions. There are 10-inch f/4 newtonians with absolutely superb optics—giving planetary images that rival a large refractor.

The perfect planetary reflector has a design used by Sir William Herschel, aptly called herschelian. In this design there are no obstructions in the light path because the observer looks directly over the edge of the tube into the primary mirror. Herschelians have huge focal ratios.

The rule of thumb should be large aperture and long focal ratio. Once again, the observer is faced with long, bulky tubes, and huge, weight-lifter-sized mounts. But reflectors are easier to adapt than refractors. Modern optical designs allow for the bending of the light path with special mirrors that eliminate obstructions and lengthen the focal ratio. The *tri-schiefspiegler* design enables a 6-inch primary with a focal ratio of f/20 to be 120 inches in focal length and to fit in a tube about 4 feet long.

There is a way to coax a longer focal length out of your newtonian. A 6-inch f/6 has a focal length of 36 inches and a large central obstruction, the secondary mirror, that reduces the image contrast. You can turn this reflecting telescope into a modified herschelian by adding an *aperture mask* (Fig. 9-3) to the system. An aperture mask is simply a piece of cardboard with a hole cut off-center. The piece of cardboard fits over the end of the telescope tube. The hole is placed so that it allows light to enter the instrument and bypass the secondary mirror and its supports. The hole is small, but the benefits are great.

For our 6-inch f/6 we will need a hole a bit larger than 2 inches in diameter. When this is placed in the system's light path, it changes the 6-inch f/6 to a 2-inch f/18 with a huge increase in the image contrast, which makes up for the loss in aperture.

Aperture masks are not limited to use on reflectors. Center the hole in the mask material and you create a mask for a refractor. You create the same effect by length-

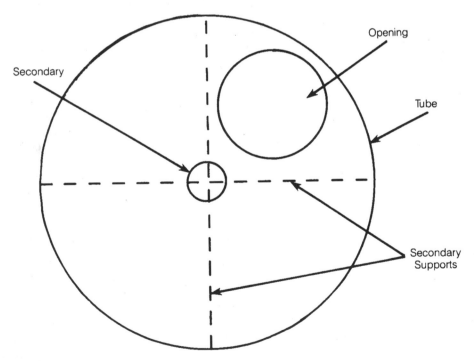

FIGURE 9-3   *The aperture mask.*

ening the focal ratio and making the aperture smaller. This can be particularly important with short-focal-ratio refractors because it helps to cut down on chromatic aberration.

An aperture mask also can help improve the overall performance of your telescope when you observe the planets. The light that reaches your eye has had a long, tough journey. Not only has it had to travel millions and maybe billions of miles through space (in the case of the planets), it has also had to make its way through Earth's atmosphere. As the light moves through the atmosphere it runs into little pockets, or cells, of air, some warm and some cold. These cells cause the image you finally see to shift and shimmer. As the light that forms your image passes through more of these cells, you will see more shimmer—and the quality of the evening's overall seeing will be poorer. If the light doesn't have to pass through too many of these cells, the image will be sharper and more stable.

A large aperture will gather light over a greater surface area, but the increased surface means you will look through more of these atmospheric cells. Using an aperture mask to reduce the light-gathering surface reduces the number of cells you are looking through and improves seeing.

Thus the aperture mask is to be used only during periods of poor seeing, not all the time. If the seeing clears up and the image sharpens for noticeable periods of time, take advantage of the full aperture; the light-gathering power and resolution of your primary will give you the best view. Use the aperture mask only as a secondary resort.

A popular telescope design is the schmidt-cassegrain (SCT). While many feel that these are the best telescopes from an overall standpoint, they do suffer problems when it comes to visual planetary observation. The primary problem is inherent in the design of the system. The mirror used in these instruments is usually a very fast f/2 in a very short tube assembly. This means that the light path is going to be reflected back to a secondary mirror a short distance away. For a 10-inch system this distance is about 2 feet. This requires the use of a large secondary mirror that blocks the light path. A large obstruction might not affect a stellar point-source image, but it can only harm the definition of a bright, extended planetary image. The SCT optical system also causes a loss of contrast because the light enters the eyepiece assembly through a hole in the mirror. This problem can be avoided by using an effective system of baffles to reduce the extraneous light that enters the eyepiece.

All these problems might seem insurmountable, but they are not. A properly baffled SCT with a secondary obstruction of less than 20 percent will give good planetary images. Just make sure that your SCT meets these simple requirements.

Then comes the latest addition to the observing family: the CCD. Should you use such technology to watch the planets? Using the CCD camera for planetary work is something most of us try after we've gotten used to our telescope—and the nighttime planetary bodies. It's usually for those times when we want to get great images of a planet—but it's something to keep in mind for the future.

But how big a telescope do you really need to observe the planets and do serious work? How big a telescope do you have? A 60-mm refractor? An 80-mm refractor? A 4.5-inch reflector? A 6-inch reflector? A 4- or 8-inch SCT? A 10-inch reflector? They are all fine, and you can make serious planetary observations with any of them. You have to remember the limitations of your particular telescope and learn to observe within them. The light grasp of a 10-inch reflector is fine until you turn it on the brilliant disk of Venus. Then you find the best view of the planet comes when you use an aperture mask, reducing the aperture to a size of 3 inches. Certain projects require larger instruments, but for every one that does there are an equal number of projects that can be successfully undertaken with smaller-aperture instruments.

OBSERVING VENUS

Venus is the easiest planet to find with the naked eye and the most difficult planet to study with the telescope. The eighteenth-century astronomer Sir John Herschel claimed this was because "the intense luster of its illuminated part dazzles the sight and exaggerates every imperfection of the telescope."

Point even a small instrument at Venus and you will see that Sir John was right. It almost hurts the eye to look at the planet. If you are using a refractor, the edges of the image will be ringed in yellow and violet, colors caused by chromatic aberration. A reflector will show light scatter from every little pit in the mirror, and the vanes of the spider holding the secondary mirror will show sharp spikes of light. Venus is indeed a very difficult object to observe with the telescope.

Any telescope will show the phases of Venus, so following the planet throughout its *apparitions*, the period of time when it or any planet is visible, is an easy project (Fig 9-4). As Venus moves in its orbit from behind and to the east of the Sun, it is in its *eastern*, or *evening*, *apparition* and will be visible in the western sky after sunset. As Venus moves higher into the western evening sky, it will eventually reach its highest point. There it is then said to be at its *greatest eastern elongation*, at 47 degrees above the horizon. It will then appear to drop back toward the horizon until it disappears in the Sun's glare. As Venus moves between the Earth and the Sun, it moves into its *western*, or *morning*, *apparition* and will be visible in the eastern sky before sunrise. Again, the planet will appear to climb in the eastern sky until it reaches its highest point at 47 degrees above the horizon, when it will be at *greatest western elongation*. Once again, Venus will drop lower in the sky until it is lost in the Sun's glare—and the entire cycle begins again.

The best time to observe Venus is when the planet has a large angular diameter and a phase large enough to show a significant portion of the disk. This usually occurs midway between western elongation and superior conjunction, and again midway between superior conjunction and eastern elongation. During these periods Venus will appear between its half-full and full phases and have a diameter of about 16 arcseconds.

We never get to see the real surface of Venus because it is shrouded by a thick, cloudy atmosphere. What you see when you observe Venus is the top layer of the planet's atmosphere. It appears to be light yellow or yellowish-white in color to the observer and almost too bright. There have been, however, tantalizing glimpses of something in the clouds over the years. In 1666 and 1667, the Italian observer Domenico Cassini saw a number of dark spots and one that was very bright on Venus. Another Italian, Bianchini, reported spots on Venus many times during the 1726–1727 apparition. Johann E. Bode, director of the Berlin Observatory, observed dark spots on the planet in 1788. In 1801, astronomer J. H. Schröter saw a dim, dusky streak. The greatest observer of the time, Sir William Herschel, was never able to make out anything more than some indistinct markings.

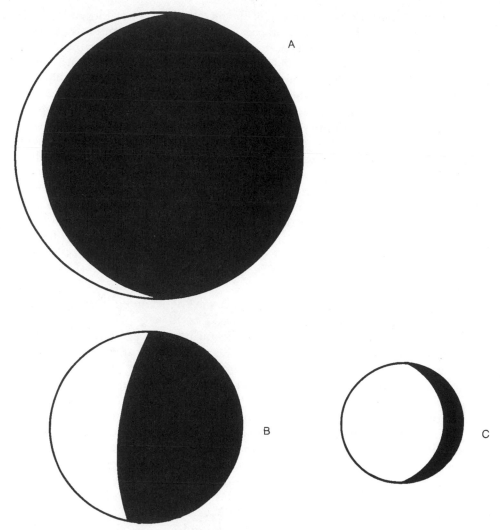

FIGURE 9-4 *Venus appears as a thin crescent of almost 1 arcminute in diameter at inferior conjunction. At elongation, Venus will appear as half-full. Near superior conjunction, it will appear as a gibbous phase almost full.*

In recent years, dusky markings that look like spokes on a wheel have been seen in the clouds of Venus. Some correlation between these ghost markings and features on the true surface of the planet have been hinted at by photographs taken at the ultraviolet end of the spectrum. And it may be true: The spacecraft Magellan mapped the surface of the cloud-shrouded planet in the 1990s—and many of the larger volcanic areas are extremely tall. This could possibly affect the Venusian clouds in those areas.

If you want to accept the challenge and hunt for these elusive features, it would be to your advantage to invest in a set of filters to help cut the glare and at least partially pierce the clouds. Observing with filters allows you to *screen* the planet in a particular wavelength of light. Because of the increased contrast induced by use of the filters, features can become more apparent. Filters act to transmit certain wavelengths of light and turn away others (Fig. 9-5). A red filter will allow red light to pass through it while it turns away light at the blue end of the spectrum. A reddish-colored feature will appear bright and stand out from the background, and blue features will blend into the disk and be difficult to distinguish.

A number of companies sell filter sets that screw onto the barrel of the eyepiece. These are useful, but every time you want to change filters you must remove the eyepiece, unscrew the filter, screw in a new filter, and then replace the eyepiece in the focuser. This can get to be a bit tedious. You can also buy filters by the sheet and hold them between the eyepiece and your eye. You don't need a large filter. Those best suited for planetary observing are the Kodak Wratten filters. They are sold in sheets 3 inches square—and they are expensive. A number of small companies offer laminated cards that have up to eight filters mounted on them. You can take the filter sheets and mount individual colors in 2 × 2 mounts or even put a series of filter colors in a single mount. Table 9-1 gives you a breakdown of the Wratten filter line, their various transmission factors, and the best filters for observing certain planets (and in some cases, the Moon).

If you cannot find Wrattens, there are other ways of obtaining filters. Theatrical supply companies often distribute sample packs of their gelatin lighting filters. Although these do not have the transmission qualities of a Wratten and might include some strange color names like Tiger Red, they can be used. The important thing in using any filter is that you know its transmission range. These theatrical filters come with a breakdown of their wavelength transmissions, so you can select the ones closest to the Wratten range.

Again, as we discussed with filters and the Moon, another useful filter is the variable-density polarizing filter. It is made of two disks of polarizing material mounted together so that either filter element can be rotated independently. When the elements are rotated so that the axes are in phase, or parallel, light transmission is at its maximum. As you rotate either element, the resulting scene darkens until the

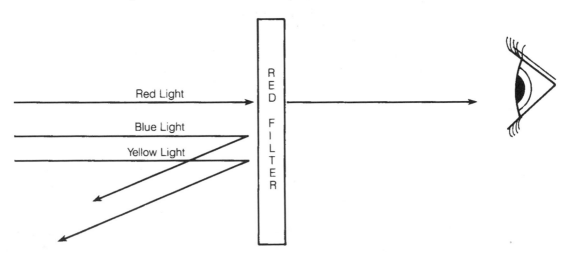

FIGURE 9-5 *Filters selectively admit certain wavelengths of light while reflecting others away.*

TABLE 9-1.   WRATTEN FILTERS OF INTEREST TO THE AMATEUR ASTRONOMER

WRATTEN NUMBER	FILTER PASSES	BEST USED TO OBSERVE
25	red	Moon, Mars, Venus
23A	red-orange	Mars
21	orange	Mars
15	deep yellow	Mars, Jupiter, Saturn
12	yellow	Mars, Jupiter, Saturn
8	light yellow	Mars, Jupiter, Saturn
57	yellow-green	Jupiter
58	green	Moon, Mars, Jupiter, Saturn
64	blue-green	Mars, Jupiter
80A	light blue	Mars, Jupiter, Saturn
82	light blue	Jupiter
82A	light blue	Jupiter, Saturn
38	light blue	Jupiter, Saturn
38A	blue	Mars, Jupiter, Saturn
47	deep blue-violet	Moon, Mars, Venus
30	light magenta	Mars

elements' axes are at a 90-degree angle to each other and very little light is transmitted. A polarizing filter allows you to cut down on the glare from an object like Venus without affecting the white light transmissions of the image.

One more filter that has come on the market speaks of our times is the light-pollution filter, sold by a multitude of companies usually for around $100. These filters block out the most common wavelengths of light pollution and in turn let in the desirable wavelengths—especially for deep-sky objects. It's only a temporary fix, of course, and it would be nicer to become active in decreasing the light pollution in your area.

When hunting for the dusky features on the disk of Venus, you can use a violet (Wratten #47) or blue (Wratten #38A) filter to bring out these low-contrast features. Because the Wratten #47 filter is so dark, you should use it only if your instrument has sufficient light-gathering power to penetrate the filter. A 6-inch telescope or larger should be used in conjunction with this filter. If you have a telescope that is smaller than 6 inches in aperture, the Wratten #38A will give adequate penetration.

In some instances, you may want to stack the filters for a better view. For example, by stacking the #58 green and #80A Medium Blue filters, you can reduce the severe twinkling of Venus—a better way of observing the changing phases of the planet.

Another area of investigation for the amateur's telescope is a study of the terminator over the course of a number of apparitions. Because Venus exhibits phases as the Moon does, it will show a terminator just like the Moon. On Venus, this line separates the light from the dark portions of the planet (Fig. 9-6).

The terminator, however, is not always a sharply defined line as it is on the Moon and exhibits a variety of conditions you can study with a telescope. For example, it can show a number of irregularities along all or part of its length. These might

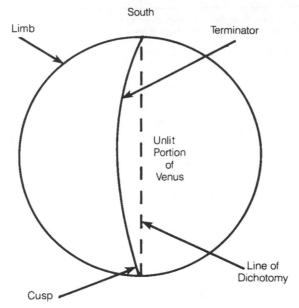

South

Limb

Terminator

Unlit
Portion
of
Venus

Line of
Dichotomy

Cusp

FIGURE 9-6   *Venus through a telescope.*

appear in white light and are best seen when using a polarizer. They also appear very well under red filtration, and a Wratten #25 will bring them out. These irregularities might look like indentations that go from the dark, unlit side of the planet into the bright, sunlit portion. They also might appear as bulges that move from the bright to the dark hemispheres of the planet. Finally, they might appear as a wavy or serrated line alternating between the two hemispheres.

Related to terminator irregularities is a phenomenon known as the *schröter effect*. As the planet moves around the Sun, it is expected to display certain phases at certain points along its orbit. When the planet is at or near either of its elongation points, it is expected to be at half phase, with the terminator theoretically being a straight line running from the north to the south pole.

The exact moment of the half phase is known as *dichotomy* and can be computed accurately to within seconds of when it is supposed to occur. The only problem is that the exact dichotomy is rarely, if ever, observed when it is supposed to take place. The difference between computed dichotomy and the observed half phase can vary by as much as a week. This is thought to be caused by refraction in the planet's atmosphere and the fact that although the limb of the planet appears sharp and distinct, the terminator shows up as a fuzzy line.

The tips of the terminator, called the *cusps* or horns, can also take on characteristics. Prominent bright areas, or *cusp caps*, have been observed at or near the true cusps of the planet during its crescent phase. They might appear to take on irregular shapes and can be seen extending along the limb of the planet. The caps have also been observed to have a dark border around them, called *cusp bands*. The cusps of the planet have also been noted to be of unequal brilliance and size, with the south cusp being the more prominent of the two. It would appear that these cusp effects are similar to the polar caps on planets like Earth and Mars. In Venus's case, however, these are not permanent features. Data from spacecraft indicate that the planet does not have features analogous to polar caps.

A final feature of the cusps is their occasional extension so that they encompass the entire planet, giving it a *halo*. Caused by the refraction of sunlight through Venus's thick atmosphere, this halo effect is most often seen when Venus is near inferior conjunction and appears as a very thin crescent.

A final observable feature on Venus is its *ashen light*. First observed by the Italian astronomer Giovanni Riccioli in 1643, the source of the ashen light has continued to elude observers even after the arrival of space probes. The phenomenon appears as a faint glow on the dark side of the planet and seems similar to *earthshine*—light reflected from the Earth and seen on the darkened Moon, especially during a lunar eclipse. However, any similarities end there. Earthshine and the ashen light might look alike, but they are caused by completely different mechanisms.

One guess for Venus's ashen light is that solar particles activate a mechanism in the atmosphere of Venus, creating a glow similar to Earth's aurora. It would seem to follow that during periods of great sunspot activity the ashen light should be visible. Although there is some evidence to show a connection between solar activity and the ashen light, there is also evidence that indicates the two are not connected. Observations by members of the British Astronomical Association (BAA) during the period of maximum solar activity in the late 1950s (1956–1958) included many sightings of the phenomenon. But observations during later periods of solar maximum do not show much in the way of observed ashen light. Also, data from spacecraft sent to Venus show no evidence for the light's existence; by the same token, however, they don't show any evidence to rule out its existence either.

The ashen light is a perfect example of negative data being important as positive data. If you observe Venus over the course of a number of apparitions, it is just as important to note when the ashen light was not seen as when it was. The only way to solve the mystery of the ashen light is to have as wide a database as possible from which to work. The only way to build such a base is to have as many observations, both negative and positive, as possible.

OBSERVING MARS

Outside of the Moon, Mars is the most fascinating object in the solar system for the man on the street. Over the centuries, a mystique has built up around the red planet. In fact, when Orson Welles broadcast his adaptation of *The War of the Worlds* in 1938, well over 1 million people were willing to believe that we were being invaded by men from Mars. More recently, spacecraft to the red planet—from the Pathfinder and its tiny roving robot, Sojourner, to the Mars Global Surveyor—have continued to pique the interest of every Earth-bound person.

A TALE OF TWO ORBITS

The red planet Mars is the first superior planet we encounter. The two most important points in a superior planet's orbit with respect to Earth are *conjunction*, when the planet is opposite Earth and moving behind the Sun, and *opposition*, when the planet is closest to the Earth in its orbit. Although superior planets are farther from the Sun than Earth, they can still show phases. When a superior planet is near conjunction at the far end of its orbit, it will show its maximum phase, slightly past gibbous, and have the shape of a 10- or 18-day-old Moon.

The orbital mechanics of a superior planet also dictate when we on Earth will get a better view of the body. Mars, however, has a funny orbit. Although the Earth reaches opposition with the red planet every 2 years, not every opposition is good for observing—mainly because its orbit is not a perfect circle (Fig. 9-7). The average opposition, which takes place every 26 months, puts the planets about 50 million miles apart and yields an apparent diameter for Mars of about 17 arcseconds. When Mars is closest to the Sun, at *perihelion*, the orbits of the two planets can be very close to each other, between 33 and 35 million miles, but because Earth moves faster in its orbit than the more distant Mars, it is not always there to greet the red planet. The most favorable views are afforded us when the planet's closest approach to Earth and the Sun coincide.

These *perihelic* oppositions occur every 15 to 17 years and take place during the late summer. When these two orbital points coincide, the planet Mars can have a disk between 23 and 25 arcseconds in diameter, and during these perihelic oppositions, the southern hemisphere of the planet is always tilted toward Earth.

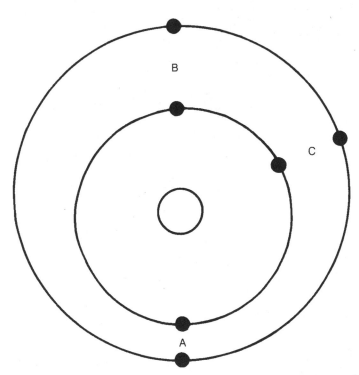

FIGURE 9-7   *Oppositions of Mars. At a perihelic opposition (a), Earth and Mars are close together. At an aphelic opposition (b) the two planets are much farther apart. The placement in (c) is when most oppositions occur.*

It is much more common for the Earth to catch up with Mars when the planet is at *aphelion*, or its farthest point from the Sun. This takes place every 9 to 12 years. During *aphelic* oppositions, the distance that separates the two planets is much greater—from 48 to 63 million miles. Because this distance is much greater than a perihelic opposition, which occurs during January and February, the apparent diameter of the Martian disk is between 13 and 15 arcseconds. During the aphelic oppositions, the northern hemisphere is always tilted toward Earth.

TELESCOPES AND MARS

As with everything else in amateur astronomy, what you do depends on the equipment you use. Just because you have a small telescope doesn't mean you can't do meaningful work and make contributions to our ever-expanding body of knowledge about Mars. You just have to temper your enthusiasm with a bit of reality. To make the best use of your instrument, it is a good idea to get and use filters. Table 9-2 lists the recommended Wratten filters for observing Mars.

If You have a 60-mm (2.4-inch) refractor, you can't expect to see minute details on the Martian surface. But during favorable perihelic oppositions, a 60-mm can easily detect the polar cap and many of the dark surface features on the planet. Although you may not be able to pinpoint very small details such as individual craters on the Martian surface, you can easily see and record the shrinking or growth of the planet's polar caps as Mars approaches opposition. Using filters you can also follow the growth of clouds near the polar caps. This might not seem like much, but observations of the polar caps contribute a great deal to our understanding of long-term Martian weather, something we are going to need a lot of information on before humans set foot on the planet. And remember, even though we have spacecraft at the planet, they do not watch for the long term—or all the planet at once.

Observing the planet Mars can be a challenge for the amateur. Because the usual apparent diameter of the planet is small, you will have to use your telescope near the limit of its useful magnification range in order to make the disk large enough to see detail. In doing so, you will have to be careful that you don't use a high magnification that causes surface features to blur.

TABLE 9-2   REACHING THE MARTIAN SURFACE WITH FILTERS

WRATTEN NUMBER	FEATURES VISIBLE
47	Clouds high in atmosphere
38A/80A	Clouds at lower altitudes
	Limb hazes
	Polar hood
64	Low-lying fogs
58	Surface frosts
8/12/15	Dust storms
23A/25	Features on surface

For example, during a favorable opposition in September 1988, Mike used his 80-mm f/11.4 refractor to observe the planet. Using the guideline of 25D, or 25 times the diameter of the objective (discussed in Chapter 4), he should have limited his maximum magnification to 75×, or 25 times 3. Images yielded by the 80-mm and a 12-mm orthoscopic eyepiece were sharp, and detail on the planet's surface was quite visible. But rules were made to be broken—especially when observing the planets. You can take advantage of particularly good seeing to raise the limit of your telescope to as much as 50 to 70 times the aperture. On a number of occasions during that apparition, Mike used an 8-mm RKE eyepiece with a 2× barlow lens to observe the planet. This yielded a magnification of 227× for his 80-mm, or 75 times the aperture (75D). The exit pupil was small, but the results were satisfactory. But this is only done when the seeing is excellent.

For most of the apparition, he used an 18-mm orthoscopic eyepiece with a 2× barlow, resulting in 101×, or about 34D. Though this is still above the limit, remember that planets take magnification much better than other celestial targets. When you increase the magnification of an image, you are making it larger without increasing its overall brightness. This spreads the available light over a larger area, and as a result the image becomes darker. Because the glare of a planetary image will wash out small details, the increased darkness of a magnified image can help bring out small surface details near the telescope's threshold of resolution.

The important thing to keep in mind is that you are trying to get the best resolution of detail possible from your telescope. There is nothing wrong with pushing your instrument to the limit when seeing permits. The key to that sentence is "when seeing permits." It is the stability of the atmosphere more than anything else that limits how far you can go. About the highest Mike pushed the 80-mm, given excellent seeing, was 227×.

A Martian Geography Lesson

Mars is the only planet whose surface features can be seen from Earth (Fig. 9-8 and Fig. 9-9). Most of the features on today's planetary charts were mapped and named during the later 1800s by the Italian astronomer Giovanni Schiaparelli. Even though Mars is only half the size of Earth, these *albedo features*, the dark markings and desert areas, cover a surface area equal to the land surface of Earth. The modern map often seen of Mars is overlaid with a longitude and latitude grid to make it easy to identify the location of each albedo feature. The longitude system for Mars begins at 0 degrees and wraps around the planet a full 360 degrees. This is to make it easier to find things on Mars when looking at the planet with a telescope.

The orientation of Mars when it is viewed through a telescope is as follows: South is at the top and north at the bottom. But the names of the east and west limbs of the planet have been changed: East is now the *preceding limb* (p) and west is the *following limb* (f). The preceding limb is the side of the planet moving into evening; the following limb is the portion of the planet under the morning Sun. Features on the planet will move west to east, from preceding to following limb, through the course of a Martian day, called a *sol*.

The terminator, also called the *phase defect*, separates the illuminated and unilluminated portions of the planet's disk. The portion of Mars illuminated by the Sun is

FIGURE 9-8 *Mars, with the Valles Marineris cutting across the planet. (NASA)*

never less than 85 percent. When the planet is seen before opposition, the terminator denotes the line of sunset on the planet and appears on the evening (p) limb. After opposition it marks the line of sunrise and appears on the morning (f) limb.

There is another line drawn on the disk of Mars, labeled the *central meridian* (CM) of the planet. It is an imaginary line that passes through both poles at a right angle to the equator and is equivalent to the prime meridian on Earth. The central meridian is used to tell us the longitude of the feature passing across the center of the planet's disk. If the planet is past opposition and a phase defect is present, the CM will appear off center on the disk.

Because a day on Mars is longer than one on Earth (24 hours 37 minutes), the planet will rotate through 350 degrees of longitude for each 24-hour Earth day. The difference of 37 minutes causes markings on the planet to appear to move backward by 10 degrees longitude from night to night. So if you observe a feature on the following limb of the planet one night, the following night it will be around the other side of the planet. It will be five or six weeks before you can see that feature again.

Information on what longitude is on the CM for any given date is found in such references as the RCAS's *Observer's Handbook*, the *Astronomical Ephemeris*, and the

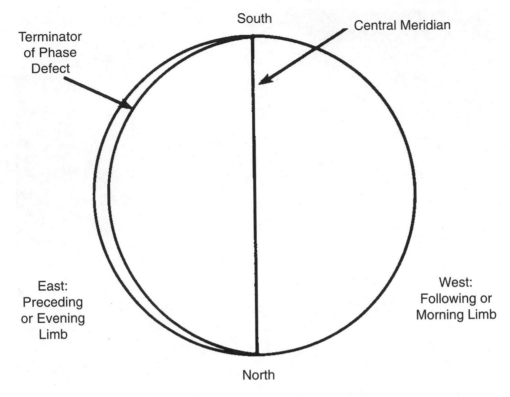

South

Central Meridian

Terminator
of Phase
Defect

East:
Preceding
or Evening
Limb

West:
Following or
Morning Limb

North

FIGURE 9-9 *Mars through the Hubble Space Telescope. (NASA)*

ALPO Solar System Ephemeris, which lists information on all the objects visible in the solar system. For example, if the ephemeris you are using gives a CM longitude of 91 degrees for a given date, you will know that the albedo feature Solis Lacus (a dark spot found at around 91 degrees on the Mars map) is on or near the center of the planet's disk.

Another piece of information that is helpful when you are observing Mars to know is the time of year on the planet. This is given in an ephemeris as the longitude of the Sun (Ls) as seen from Mars. These longitudes denote the season on the planet and may seem tough to understand at first. The Ls for a particular date is figured for the northern hemisphere of Mars. Ls = 0 degrees indicates the point for the beginning of Martian spring. Ls = 90 degrees is the beginning of summer. Ls = 180 degrees is the beginning of autumn, and Ls = 270 degrees is the beginning of the Martian winter. Perihelion always comes at Ls = 250 degrees and aphelion at Ls = 70 degrees. Remember, these apply to the northern hemisphere of the planet, so the seasons in the southern hemisphere are exactly opposite, just like on Earth. If your ephemeris tells you that the longitude of the Sun (Ls) for the date of your observation is 200 degrees and it is an aphelic opposition, you know that it is early autumn. For the same Ls during a perihelic opposition, it is early spring, in the Martian Southern hemisphere.

MARTIAN POLAR CAPS

The first thing you see when you look at Mars with a telescope is the polar cap. During aphelic oppositions the north polar cap (NPC) is prominent; during perihelic oppositions the south polar cap (SPC) is visible. Made up of frozen carbon dioxide and water ice, the polar caps are the prime movers of Martian weather. Clouds that form over the surface of the planet are fed by the evaporation or *sublimation* of the caps, particularly the larger SPC (Fig. 9-10, 9-11, and 9-12).

FIGURE 9-10 *Mars through the Hubble Space Telescope: The Valles Marineris region, centered on about 60 degrees longitude (the crescent-shaped cloud just right of center is around Olympus Mons, one of the largest, if not the largest, volcano in the solar system). (NASA)*

FIGURE 9-11 *Mars through the Hubble Space Telescope: The Tharsis region, centered on about 160 degrees longitude (Valles Marineris is at the lower left). (NASA)*

Because the polar caps are so prominent and their growth and retreat are so important to science, monitoring them on a nightly basis is a perfect project for the amateur with a small telescope. A feature on the NPC is the *polar hood*. Although the SPC can also have a polar hood, the one that forms over the NPC is more prominent. The north polar hood (NPH) forms during late spring on Mars (Ls = 70 to 80 degrees) and is associated with what scientists refer to as an *aphelic chill*, during which time the retreat of the NPC suddenly stops. Polar hoods can appear bright, so bright that you could confuse the hood for the polar cap itself. This is when filters come in handy: To determine if you are seeing a polar hood or the polar cap itself, look at the feature first with a blue (Wratten #38A) filter. If the feature is a cloud or the hood, it will appear bright in the blue light transmitted by the filter. Quickly switching to a red (Wratten #25) filter will cause the feature, it if is the hood, to disappear.

FIGURE 9-12 *Mars through the Hubble Space Telescope: The Syrtis Major region, centered on about 270 degrees longitude (the dark "shark fin" feature is Syrtis Major, with the impact basin Hellas below it). (NASA)ASA)*

With larger instruments, over 8 inches in aperture, you may be able to follow the shrinking of the polar cap and watch features the retreating polar cap uncovers. For example, in the NPC, look for small projections of ice ringing the NPC during spring and summer. These are known as *Ierne* at longitude 150 degrees, *Lemuria* at longitude 60 degrees, and *Cecropia* at longitude 270 degrees. These features are best seen using red and green (Wratten #58) filters. Features in the SPC to watch for include the ice projections called *Novus Mons* at longitude 310 to 330 degrees, *Thyles Collis* at longitude 230 degrees, and *Thyles Mons* at longitude 180 degrees. These features are best observed during mid-spring (Ls = 230–240 degrees) and disappear with the full arrival of summer in the southern hemisphere. You also may notice that during summer, the SPC is divided into two sections separated by the Mountains of Mitchell, which are not mountains but another ice remnant of the fully developed polar cap.

MARTIAN WEATHER

For a planet with so little available liquids to condense and form clouds, Mars is very active meteorologically. During favorable oppositions, even the owners of small telescopes can follow the development of clouds and dust storms across the surface of the planet. Filters are very important for the observer wishing to study the Martian clouds because they can tell the observer just how high a cloud is in the planet's atmosphere. Violet (Wratten #47) filters penetrate the Martian atmosphere the least; this filter is useful in detecting high-altitude cloud features. Blue filters (Wratten #80A, 38, and 38A) penetrate farther into the atmosphere and enhance clouds at lower levels than the violet filter. Blue-green filters (Wratten #64) emphasize features low to the ground but still elevated. A green (Wratten #58) filter will brighten features such as frosts on the surface of the planet.

The easiest clouds to detect on Mars are those forming along the slopes of highly elevated areas like the huge volcano Olympus Mons. These are called *orographic clouds* which appear as single, isolated bright spots in the Martian atmosphere when viewed without a filter. They are best seen in the spring and summer when the polar cap is shrinking. When viewed through blue and violet filters, they form and move through the upper levels of the planet's atmosphere. The wispy clouds that form near the peaks of mountains and the tops of tall buildings like the Sears Tower are Earth's counterparts to orographic clouds on Mars.

Another target for the small telescope is the detection of *limb hazes* along the terminator. These atmospheric features do not rotate with the planet—thus they are probably localized cloud formations caused by rapid changes in temperature at Martian nightfall or before sunrise. They extend high into the atmosphere (as indicated by their brightness when seen through blue and violet filters), and they can be harbingers of upcoming weather changes on the planet. The color of these limb hazes can also indicate if a dust storm has begun on the side of the planet facing away from Earth.

Because Mars is subject to rapid and drastic temperature variations, *frosts* and *fogs* appear in low-lying surface areas. Visible as bright spots or patches, usually in the planet's desert regions, they are deposits of frozen carbon dioxide and water-ice laid down during the cold Martian night. As these frosts are heated by the Sun, they create fogs. By local Martian noon, they have usually disappeared. Because the frost is deposited directly on the surface, it will appear bright when viewed through a yellow or green filter. The fogs hang above the surface and are best seen with a blue-green filter.

The bane of Mars observers has always been dust storms. Growing swiftly from the desert basins, these storms can quickly blanket large areas, or even the entire planet, in a matter of days. Since 1877, more than 40 dust storms of varying intensity have interrupted visual observations from Earth. Six of these (four of them in the last three or four decades) have been so severe that all sight of the planet's surface was lost. Dust storms are caused by winds so severe that they make an earthly tornado seem like a spring zephyr. Racing across the Martian landscape at nearly 300 miles an hour, the dust storm throws sand and surface material miles into the atmosphere. While the storms arise suddenly and can stop abruptly, the materials they have tossed into the atmosphere may take weeks or months to filter back down to the surface.

Yellow clouds, usually seen in the area of Solis Lacus, Chryse, and other basin areas, are often precursors of these dust storms. They show up best with yellow filters;

orange (Wratten #21) or red filters will show the observer the cloud's boundaries. Although a dust storm can start at any time during the Martian year, the planetwide storms observed since 1970 have all begun when the longitude of the Sun was between 200 and 360 degrees—early to late summer for the southern apparitions, and early to late winter for the north.

MARTIAN SURFACE FEATURES

After a first-time observer has seen the Martian polar caps, the next features to catch the eye are the patches of green that dot the disk. These are the raised areas of the planet. They appear quite distinct from the depressed, orange areas defining the planet's deserts and basins. These albedo features seem to darken with the retreat of the polar cap. The changes do not take place because of the growth of vegetation spurred by the water from the poles—as once thought. The areas surrounding the dark features become lighter as dust and particles suspended in the atmosphere during the winter filter down to the surface. The darkening of the albedo features is nothing more than an effect of contrast.

The albedo features can undergo other changes throughout the course of a Martian year. The region of Hellas, a huge impact basin, undergoes weather-related changes during the southern hemisphere's late autumn and early winter, and should be watched closely. A gray hood often covers the area until the region moves into spring. As the planet moves into southern spring, the overall weather activity of the southern hemisphere increases: With the retreat of the SPC, Hellas will undergo changes such as the growth of frost patches in the southern area of the basin, and cloud activity moves up from the SPC. Later in spring, the region should clear of ice and frost and its floor will appear to darken in color. As summer gets closer, the basin shows signs of dust storms, which should be closely monitored by the observer. In fact, three dust storms, two during the 1986 opposition and one during the 1988 opposition, began in the Hellas basin.

Changes like those in Hellas are seasonal, last only a short time, and can be predicted with some accuracy. Areas of the planet also can undergo longer-lasting *secular* changes that can occur abruptly and alter the appearance of an albedo feature for a long time.

Solis Lacus and Syrtis Major are particularly prominent features on the planet that can easily be monitored. When viewed in a telescope, Solis Lacus appears as a dark feature surrounded by a lighter desert area. It looks like a giant eye staring out from the surface of Mars, which is how it got its common name, "the Eye of Mars." Observers of the area report that the Eye of Mars can undergo dramatic changes in its shape, size, and brightness. During recent oppositions it has reportedly grown larger and darker while very delicate, dark lines appeared in the surrounding areas.

Syrtis Major is probably the most easily recognized feature on the planet. First seen by Christian Huygens in 1659, observations of the feature made during the next few years allowed early astronomers to determine that the planet rotated on its axis in 24 hours 40 minutes—a difference of 3 minutes from today's rate as determined by the Viking orbiter. Syrtis appears as a dark, triangular feature surrounded by lighter desert areas. Seasonal changes are predicted by the ALPO Mars Section for the peninsula-like feature each time the planet passes perihelion, when it seems to narrow, and after aphelion, when a general expansion of Syrtis takes place. The area has also been subject to a number of secular changes over past apparitions. During the past 15 or so oppositions, Syrtis Major has appeared to become smaller in size, losing the sharp "parrot-beak" at its northern tip and growing more narrow across its center. This could indicate the expansion of the desert areas, which would ultimately change the entire face of Mars.

Occasionally visible in larger telescopes are the spidery, almost gossamer-like markings of the famous canals of Mars. First observed by Schiaparelli in 1877, these features have stirred up more controversy than any other feature on Mars. On the maps Schiaparelli crafted during the 1877 opposition, he drew a number of dark, almost straight lines that he labeled *canali*. To Schiaparelli, who wrote in Italian, this was the word for *channel* and was meant to indicate a long, almost straight feature that had nothing to do with intelligent beings. The problem began when other astronomers who did not speak Italian wrongly translated "canali" to "canal." When the public heard about this, there was chaos—there had to be intelligent beings on Mars, digging canals.

The debate raged for 17 years. Then an east coast amateur astronomer named Percival Lowell entered the fray. In 1894, Lowell erected an observatory in Flagstaff, Arizona, and obtained a 24-inch refractor from the firm of Alvan Clark & Sons—a scope said to be one of their finest optical efforts. From this observatory, which still bears his name, Lowell set out to prove that Schiaparelli's canali were really the work of a race of master builders. Lowell observed over 400 canals criss-crossing the planet in a complex network. His efforts to popularize his theories met with some success—and he was even able to convince some prominent astronomers that he was right about the canals.

The arrival of spacecraft on Mars settled the argument that raged through astronomical and popular literature for almost 100 years. The spacecraft to Mars, the Mariners, found no canals; and the Viking landers were not met by members of Lowell's race of master builders. But, strangely, the spacecraft did confirm that Lowell's observations were not imaginary. Some of his canals appeared to be chains of dark-floored craters, and some followed closely the alignment of dark-surface patches on the planet.

The discovery of Valley Marineris showed that Lowell did see something big on the red planet. The "Valley of the Mariners" was discovered by the Mariner missions to Mars in the late 1960s. It stretches over 3,000 miles, is 150 miles wide and 4 miles deep, and is often referred to as the "Grand Canyon of Mars." It is a huge canyon that crosses one quarter of the surface, looking from above as if someone had slashed through half the planet. Theories abound as to how the valley was formed. Some scientists say it was cut by furious water action at some time in the planet's distant past. But most scientists believe it was caused by the movement of the planet: a huge fault line in the surface that collapsed after a major volcanic eruption.

If you take your telescope out during the next favorable opposition and seek out the feature Tithonius Lacus, you might notice a network of bright patches of fog and mist in the area. Fog and mist gather in the Valley of the Mariners and are easily visible from Earth, seeing permitting.

The valley is evidence that of the two, Schiaparelli and Lowell, Schiaparelli was closer to the truth. The valley is a "canal" in the true sense of the word: a channel. But what did Lowell see? Did he waste his fortune, time, and reputation pursuing an optical illusion? In a sense he did, because there are a number of optical effects that cause you to see canals. The eye can blend separate details into a single continuous image, appearing as a straight line. The action is very similar to the way letters are created by a dot-matrix printer. From a distance each letter appears to be a solid, continuous impression on the paper. But up close, it is easy to see the individual dots combine to create the letter. This blending of discrete images is seen by numerous observers.

The prominent astronomical observer Edward E. Barnard, possessor of the best eyes in astronomy of his time, was able to discern several craters that looked like dark patches on the Martian surface during the opposition of 1892. He used the 36-inch refractor at Lick Observatory. Are the craters that dot the surface of the planet the reality behind Lowell's canals? Go out and have a look at Mars. Then make up your mind.

If you approach the observation of Mars thinking only of seeing the legendary "canals," you are going to be disappointed. You cannot expect to see this effect every time you turn your telescope on Mars. It takes extremely stable seeing and patience to discern the subtle markings. For example, at an observing session during the excellent 1988 opposition, a number of observers commented that they had seen the "canals" using a 10-inch, f/20 tri-schiefspiegler and a 10-inch, f/9 newtonian. These observers, including ALPO Assistant Mars Recorder Dan Troiani, were blessed with spectacular seeing conditions. The air was saturated with moisture which placed some estimates of the seeing conditions at 10, the highest level (Fig. 9-13 and 9-14).

DRAWING MARS

One of the real joys of amateur astronomy is sharing the results of your observations with your family, friends, and other observers. Your observing logs, journals, diaries, or whatever you call them, are one way to do this. In the case of an object like Mars, however, they require you to reduce visual impressions to words—and that is often difficult to do. There is often so much detail visible on the planet that words cannot convey their beauty and power. As someone once said, "A picture is worth a thousand words." That person must have been looking at Mars.

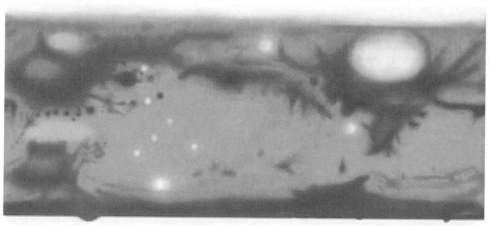

FIGURE 9-13 *A strip drawing of Mars. (Courtesy of Dan Troiani)*

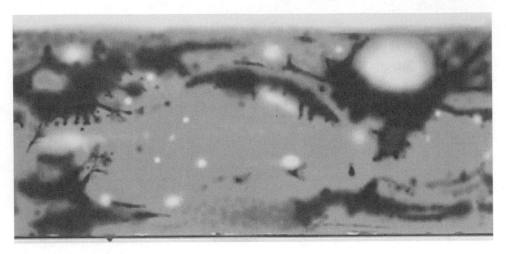

FIGURE 9-14 *A strip drawing of Mars. (Courtesy of Dan Troiani)*

Making astronomical drawings is not as difficult as it seems. When drawing a planet like Mars, you are not trying to rival Van Gogh. Very few people have that talent. What you are doing is making a permanent record of your visual impressions of the planet's surface over a short period of time. "You are not making a work of art," says Dan Troiani, "you're making a work of science." In Chapter 8, we learned a simple way to make accurate renditions of lunar features using outline charts. When you draw the surface of Mars you are dealing with an ever-changing subject, so you won't have the luxury of using a static start provided by an outline chart.

The first thing to do is draw a blank circle. It can be drawn on a page in your journal, a 4 by 6 card, or a sheet of paper as long as it meets one requirement: It must be the size required by the organization to whom you will eventually submit the drawing. Most groups require you to use a blank disk between 40 and 50 mm across. The Association of Lunar and Planetary Observers (ALPO)—a group that gives you a logical reason to make and contribute planetary drawings—can supply you with forms for recording images on Mars and the other planets (Fig. 9-15 and 9-16).

Using a compass or a template of the proper size, draw a few disks on file cards for use at the telescope. You will also need a few pencils, sharpened and of varying hardness, a timepiece that keeps accurate time, a red-shielded light, and something to work on such as a simple clipboard. Before you head outside, take down your ephemeris and check the night's phase defect (terminator). Take a couple of minutes to draw it in on a few of the blank circles. If the planet shows a phase defect and you don't record it on the forms, they will be worthless—because every detail you record will be in the wrong place.

Now go out to the telescope and begin by getting comfortable at the eyepiece. If you are using a refractor, pull up a chair and sit down if you can. If you are using a reflector, get as comfortable as possible at the eyepiece. One advantage of long-focal-length newtonians is that you will probably have to use a ladder to reach the eyepiece, and the ladder will provide you with a convenient ledge on which to rest your supplies. Select an eyepiece that allows you to see as much detail as possible without

A.L.P.O. Mars Section Observation

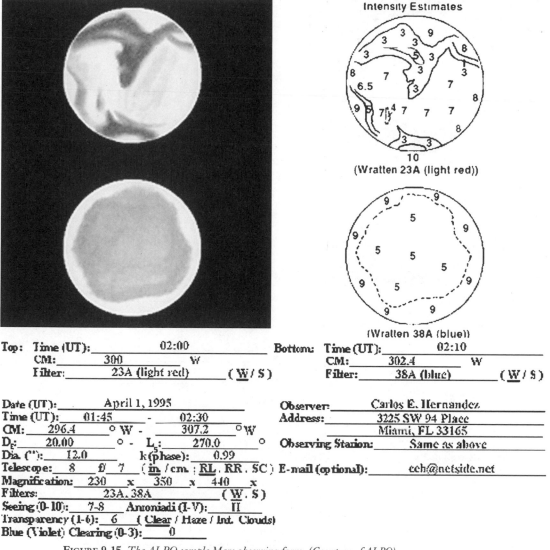

Intensity Estimates

10
(Wratten 23A (light red))

(Wratten 38A (blue))

Top: Time (UT):	02:00		Bottom: Time (UT):	02:10	
CM:	300	W	CM:	302.4	W
Filter:	23A (light red)	(W / S)	Filter:	38A (blue)	(W / S)

Date (UT):   April 1, 1995

Time (UT):   01:45   -   02:30

CM: 296.4   ° W -   307.2   °W

D$_E$:   20.00   ° - L$_s$:   270.0   °

Dia. ("):   12.0   k (phase):   0.99

Telescope: 8   f/ 7   (in. / cm. : RL . RR . SC)

Magnification: 230   x   350   x   440   x

Filters: 23A. 38A   (W . S)

Seeing (0-10):   7-8   Antoniadi (I-V):   II

Transparency (1-6):   6   (Clear / Haze / Int. Clouds)

Blue (Violet) Clearing (0-3):   0

Observer: Carlos E. Hernandez

Address: 3225 SW 94 Place

          Miami, FL 33165

Observing Station: Same as above

E-mail (optional):   ceh@netside.net

FIGURE 9-15 *The ALPO sample Mars observing form. (Courtesy of ALPO)*

fuzzing out the planet. Record your time and give the image a chance to settle down. Take a few minutes to reacquaint yourself with the visible features. Try various filters to see what effect they have on seeing. Watch the focus of the image. Changes in seeing may require you to refocus the image from time to time.

Begin your actual drawing session by noting on your form the date and time you start. Because Mars rotates at about the same rate as Earth, you will have plenty of time to get all the details on your sketch. Estimate the overall seeing on a scale of 0 to 10 and the transparency on a scale of 0 to 6. Draw in the polar caps, south at the top

A.L.P.O. Mars Section Observation

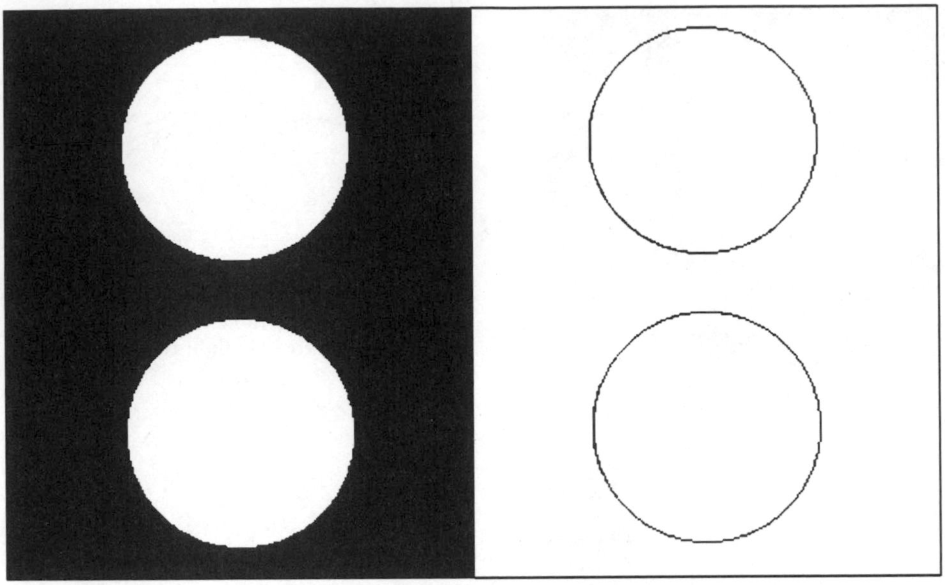

Top: Time (UT):_____   Bottom: Time (UT):_____
     CM:_____° W   CM:_____° W
     Filter:_____ (W / S)   Filter:_____ (W / S)

Date (UT):_____   Observer:_____
Time (UT):_____ - _____   Address:_____
CM: _____° W - _____° W
D_E:_____° - L_s:_____°   Observing Station:_____
Dia ("):_____   k (phase): _____
Telescope:_____ f/_____ (in. / cm. ; RL . RR . SC)   E-mail (optional):_____
Magnification:_____x_____x_____x
Filters:_____ (W / S)
Seeing (0-10):_____ Antoniadi (I-V): _____
Transparency (1-6):_____(Clear / Haze / Int. Clouds)
Blue (Violet) Clearing (0-3):_____

Notes

_____
_____
_____
_____
_____
_____
_____

FIGURE 9-16 *The blank ALPO Mars observing form. (Courtesy of ALPO)*

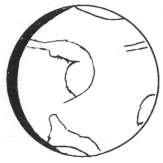
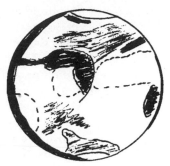

FIGURE 9-17 *Drawing Mars I.*   FIGURE 9-18 *Drawing Mars II.*   FIGURE 9-19 *Drawing Mars III.*

and north at the bottom, depending on whether one or both are visible (Fig. 9-17). Go over the telescopic image and study it in white light (with no filter), and then with a yellow or red filter to identify surface features.

Start at the center of the disk and draw an outline of the major features (Fig. 9-18). These steps should take about 15 minutes to complete. Now grab your filters and start to add details to those outline forms. Darken the familiar features, and indicate surface details and atmospheric features with dotted outlines for bright features like frost, and dashed outlines for clouds and hazes (Fig. 9-19).

The entire process of drawing the Martian disk should take from 40 to 50 minutes. You can't take any longer because the details have to be recorded before they rotate around the limb of the planet. It's not a bad idea to get the detail on the preceding limb down first, because that will be the first area to rotate out of view. When you have completed the drawing, give the planet a final once over and note the time of completion on your form.

Back inside, take a few minutes to add information on the telescope and magnifications you used during the drawing session, the filters you used (if they are not Wrattens, try to include their transmission factors), and the longitude of the CM at the beginning and end of the session as obtained from an ephemeris. While you have the ephemeris open, scribble down the Ls for the date of the drawings. If you are going to submit this to ALPO, transfer the information from your journal or record card to their report form. Even if you don't submit your Martian information to the ALPO, such record keeping is good practice—and a good habit to get into.

OBSERVING JUPITER

If you liked the Martian show, with its changing weather patterns and wide variety of surface detail, you are going to love Jupiter. You can contribute a great deal of important work studying Mars with a small telescope, but Jupiter can be a veritable treasure house for the amateur.

The biggest advantage Jupiter gives the amateur—even with a 60-mm telescope—is its size. At its smallest, Jupiter is 30 arcseconds in diameter, larger than Mars at its best, at its largest apparent diameter, 50 arcseconds, it is absolutely spectacular. Its magnitude range, between −1.3 and −2.7 with an average at opposition of −2.3, makes it easy to find with the naked eye. Unlike the orbit of Mars, with its great eccentricity, the nearly circular orbit of Jupiter makes it easily visible for 10 out of every 13 months. Like Mars, Jupiter will show a very slight phase defect when it is near conjunction. For all intents and purposes, Jupiter was tailor-made for the amateur astronomer.

Unlike Mars, Jupiter does not have a visible solid surface. The "surface" we see is actually the top of a great ball of gas, Jupiter's atmosphere. Constantly in motion, the Jovian atmosphere rotates at two different speeds. The equatorial zone and its accompanying belts, System I, rotates in 9 hours 50 minutes; the rest of the planet, System II, takes 5 minutes longer to circle the planet. The features appearing across this rapidly moving stage are the tops of clouds and storm systems—all driven by winds moving at greater than hurricane force.

Even a 60-mm refractor can show an amazing amount of detail on Jupiter. At 100× you will easily see its squashed appearance due to its rapid rotation and the two, or more, major belts of the planet. The Great Red Spot (GRS) is also visible. If it is in a period of dormancy, you might be able to detect the great cavity in which it resides, the Great Red Spot Hollow (GRSH).

Details on the planet are not the only attractions. With its retinue of four bright satellites, the Jovian system offers the small telescope user a variety of satellite phenomena to observe. As they move in their orbits around Jupiter, the satellites cross the face of the planet (transit), move behind the planet (occultation), and move through the shadow of the planet (eclipse). In addition, the shadows of the satellites can be seen transiting the planet. This is a very busy system to observe.

For the observer with a larger telescope, fine details are to be seen in the belts of Jupiter. Watching and recording the movements of tiny light and dark spots and splotches across the surface of this giant planet can give you a lifetime of pleasure and the opportunity for scientific contribution.

TRANSITS OF THE CENTRAL MERIDIAN

Geography on Jupiter is defined by the systems of light areas, or *zones*, and dark areas, or *belts*, that circle the planet (Fig. 9-20). The closest thing to a permanent feature on the planet is the GRS, which has been around for over 300 years, and three white oval—and often changing—spots visible in the south temperate belt (STB) since 1939.

This lack of permanent features and the planet's rapid rotation make it difficult to determine the location of any features on Jupiter. As with Mars, the primary resource for locating, and relocating, a feature on Jupiter is by the longitude of the feature. How can you determine the longitude of something moving—with no points of reference in the background? This is not the impossible task it seems. All you need is a telescope (from 60-mm on up), pencil, notebook, accurate watch, calculator, and copy of this year's ephemerides. The reduction of your results are so simple that the calculator is optional.

The basic work has already been done for you by the wonderful people who publish your ephemeris. All you have to do is note the time the feature passes the

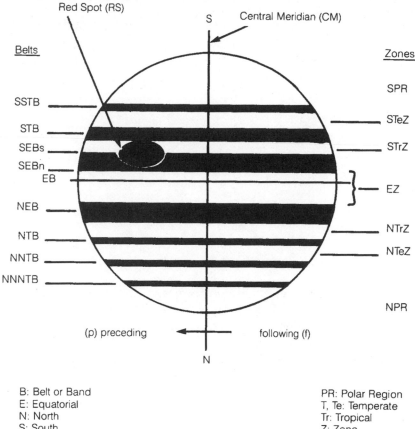

Red Spot (RS)   S   Central Meridian (CM)

Belts   Zones

SPR

SSTB   STeZ

STB   STrZ

SEBs

SEBn

EB   EZ

NEB

NTB   NTrZ

NNTB   NTeZ

NNNTB

NPR

(p) preceding   following (f)

N

B: Belt or Band   PR: Polar Region
E: Equatorial   T, Te: Temperate
N: North   Tr: Tropical
S: South   Z: Zone

FIGURE 9-20 *Jupiter's belts and zones.*

planet's central meridian (CM). The CM of Jupiter is the same as it was on Mars, an imaginary line running from pole to pole at a right angle to the equator. On Jupiter, determining the CM is a bit easier than it is for Mars because Jupiter is not tilted toward or away from Earth by more than a few degrees. You can eyeball the CM or purchase a *micrometer reticle eyepiece* (MRE).

An MRE has two or more finely scribed lines at right angles to each other, etched on a glass plate that is placed at the focal plane of the eyepiece. To use it, simply position one of the lines parallel to Jupiter's equator, which is roughly in the middle of the equatorial belt (EB). The other line defines the planet's CM. When the feature you are observing crosses that line, determined by sight or by the reticle, it is said to transit the CM. Because Jupiter rotates so rapidly, it is best to mark the time of this transit at three points, then average the three. The first timing is made when the leading, or preceding (P) edge of the feature crosses the CM. The second is made when the approximate center (C) crosses, and the final timing is made when the following (F) edge reaches the CM.

TABLE 9-3. ADJUSTMENTS FOR JUPITER CM TRANSITS

MINUTES	SYSTEM I	SYSTEM II
1	0.6	0.6
2	1.2	1.2
3	1.8	1.8
4	2.4	2.4
5	3.0	3.0
6	3.7	3.6
7	4.3	4.2
8	4.9	4.8
9	5.5	5.4
10	6.1	6.0
60 (1 hour)	36.58	36.26

Source: ALPO.

An ephemeris such as the *Astronomical Almanac* lists the line of longitude on the CM of Jupiter at 0 hours UT for each day of the year. If you have not made your timings in UT, convert them by using the table in Chapter 6. Multiply the number of hours since 0 hour UT by 36.58 for features in System I or by 36.26 for those in System II. Then find the number of minutes that enter into the timing from Table 9-3 and add it all together. The result will give you an accurate determination of the feature's longitude to within 1 degree.

For example, let's say you see a dark spot in the EB on August 1, 1998, and time its three points of transit as follows:

Preceding edge (P)	5:27 UT
Center (C)	5:30 UT
Following edge (F)	5:34 UT
Average for transit	5:30 UT

From the ephemeris you determine that the longitude on the CM for 0 hours UT on that date is 75.3 degrees. Multiplying 5 by 36.58 gives you 182.9. Table 9-3 tells you that the correction for 10 minutes in System I is 6.1 degrees, so multiply this by 3 (3 × 10 = 30 minutes) to obtain 18.3. Add the two results to the listing from the ephemeris and you have a longitude of 75.3 + 182.9 + 18.3 = 276.5 degrees. If you cannot time the transit of the complete feature because it has passed the CM by the time you see it, make sure to note the section of the feature from which the timing is made (Fig. 9-21).

In addition to taking actual timings of the transit of a Jovian feature, you have to be able to describe it. Features on Jupiter are either bright and white (W) or dark (D). They can be further described by using the classifications as seen in Table 9-4; these include six dark features and six bright features. And ALPO's Web site on the Internet will also have suggestions as to how to record the various features.

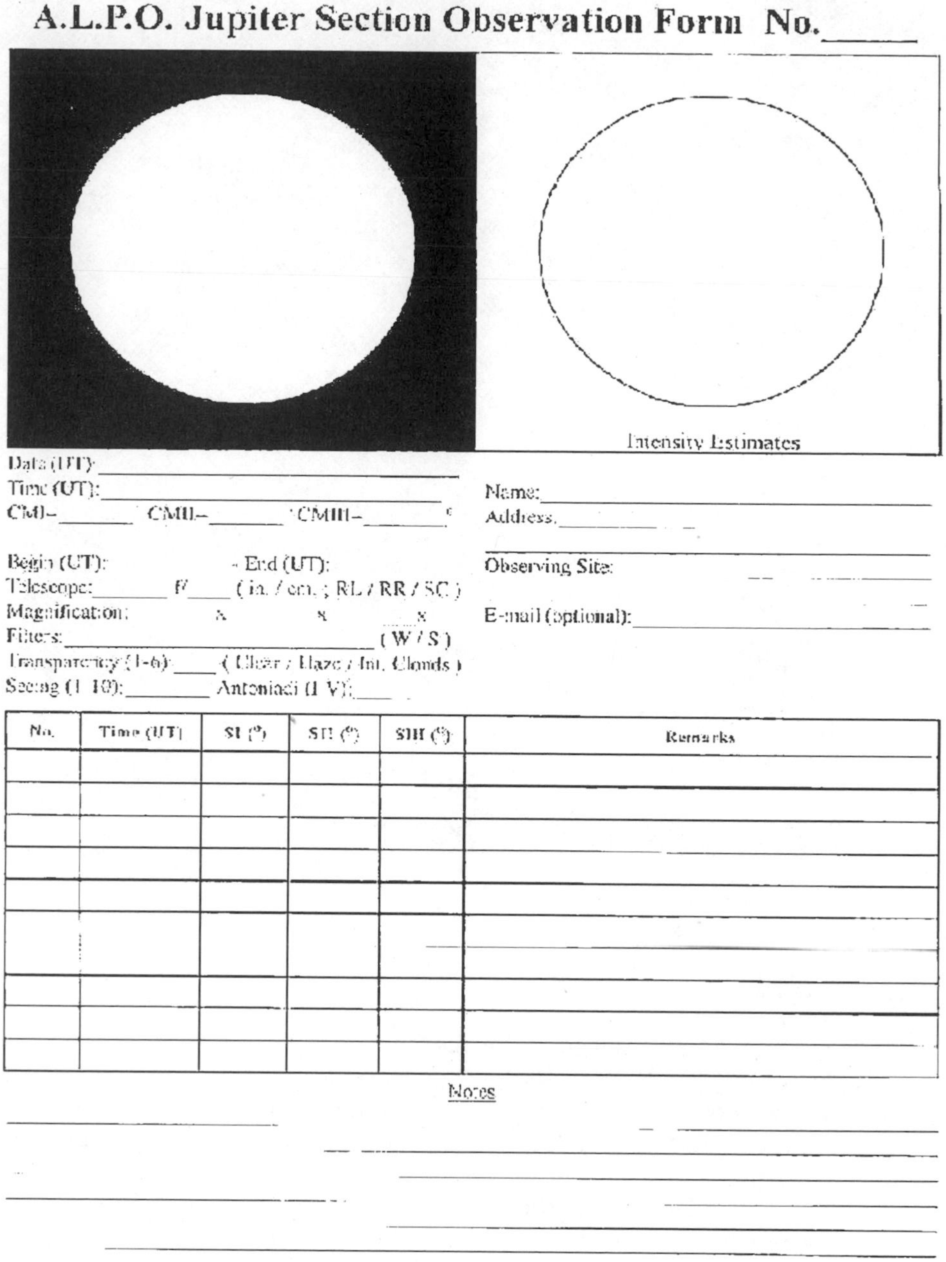

A.L.P.O. Jupiter Section Observation Form No._____

Intensity Estimates

Date (UT):_____

Time (UT):_____

CMI-_____ CMII-_____ CMIII-_____ °

Begin (UT):_____ - End (UT):_____

Telescope:_____ f/____ (in. / cm. ; RL / RR / SC)

Magnification:_____ x _____ x _____ x

Filters:_____ (W / S)

Transparency (1-6):_____ (Clear / Haze / Int. Clouds)

Seeing (1-10):_____ Antoniadi (I-V):_____ .

Name:_____

Address:_____

Observing Site:_____

E-mail (optional):_____

No.	Time (UT)	SI (°)	SII (°)	SIII (°)	Remarks

Notes

FIGURE 9-21 *The blank ALPO Jupiter observing chart. (Courtesy of ALPO)*

Figure 9-22 *The Great Red Spot on Jupiter taken by the Voyager 1, in 1979. (NASA)*

JOVIAN FEATURES

The oldest recorded feature on Jupiter is the GRS (Fig. 9-22). First observed by the English scientist Robert Hooke in 1664, it is a huge storm spinning through Jupiter's southern hemisphere. Like a hurricane in the southern hemisphere of Earth, the GRS spins counterclockwise, and it takes 6 days to rotate around its center. The GRS is lodged in a huge bay, the GRSH, that extends over the southern equatorial belt–south (SEBs) and spills into the southern tropical zone (STrZ). Winds in the belts and zones of Jupiter travel in opposite directions, adding rotation and extending the life of a feature trapped between them like the GRS.

The color of the GRS can vary from pale yellow-gray to deep red. The intensity of the colors can vary greatly from one apparition to the next. During the 1960s and early 1970s, the GRS was a prominent red-orange color, but it began to fade in 1976. Observers in the last decade noticed an increase in the prominence of the GRS, suggesting that the deep-red color, which has been its trademark, is returning. If the GRS is not visible, an observer may be able to detect its bay, the GRSH, a hollow and

TABLE 9-4. JOVIAN DESCRIPTION CLASSIFICATIONS

DARK (D) FEATURES

Projections: An area extending from the edge of a belt.
Rod: An elongation running parallel to the equator.
Festoon: A filament crossing a zone.
Loop festoon: A curved festoon extending from a projection.
Column: A vertical form penetrating from a belt to a zone.
Disturbance: Any large, sharply defined area.

BRIGHT OR WHITE (W) FEATURES

Oval: A bright, well-defined round area.
Nodule: A feature similar to but smaller than an oval.
Notch: A semicircular indentation on the edge of a belt.
Bay: A feature similar to but larger than a notch.
Rift: A long, thin streak along the interior of a belt.
Streak: An elongated white spot.

---

much paler feature. If the observer looks carefully, the outline of the GRS can be made out inside the hollow.

Because of the size and prominence of the GRS, a study of the feature and its movement across the planet is ideal for the amateur with a small telescope. Transit timings taken at all three points—preceding, center, and following—can provide valuable information for determining the details of the storm.

Other features inhabiting the cloudy surface of Jupiter are not as long-lived as the GRS. Three white ovals in the STB have been observed since 1939, making them the second oldest features on the planet. Named BC, DE, and FA, they are similar to the GRS in that they spin counterclockwise and are considered high-pressure centers. Since 1980, the white ovals have been growing smaller with an accompanying increase in their brightness. It is estimated that they are in the last few years of their existence, so a careful eye should be kept on their sizes, intensities, and transits. For example, in 1983, Jose Olivarez, then an ALPO Jupiter Recorder, discovered 12 features along the south edge of the NEB. These blue spots with accompanying festoons seemed to be holes in the Jovian atmosphere. You may be able to spot such features, too—but because they are hard to discern against the disk, use a red filter to bring them out.

In addition to the spots and ovals on the planet, the belt systems are home to a wide variety of activity, primarily *disturbances* in the STrZ and the SEB. At the beginning of this century, from 1901 to 1939, the STrZ was home to a major disturbance that spread halfway across the planet. From 1939 until the mid-1970s, the dark feature seemed to disappear, only to become quite prominent again in 1978. The disturbance seems to show a definite trend toward periodical flare-ups, so the area should be kept under observation. In 1985, a white spot detected in the SEB grew to become a great white streak across the belt. The SEB bright streak disturbance has reestablished itself during subsequent oppositions.

Filters are a great asset for the observer studying the various features on Jupiter. The colors on the planet range from the brownish belts to the bright, off-white zones. The contrast between belts and zones can be increased with a light blue Wratten 80A filter: Festoons and columns exhibit a bluish cast, which is darkened by using a yellow filter.

DRAWING JUPITER

It's time to get out the pencil and paper again. Drawing Jupiter is different from drawing Mars, and the primary drawback is Jupiter's rapid rotation. Because of this factor, drawings must be executed rather quickly, in no more than 10 minutes. After this length of time, the features on the disk will have noticeably shifted position and the accuracy of the drawing will be destroyed. Drawings of Jupiter fall into two types: full-disk drawings, like those done for Mars, showing the entire planet visible at the time of the drawing; and strip sketches, used to portray details in a small band of latitude on the planet. Both types of drawings make important contributions to our growing knowledge of the giant planet, so let's begin with the full-disk technique.

As with Mars, the first thing to do is prepare a blank. Because Jupiter rotates so fast, it is not a perfect circle. The equator bulges out and is seen as being wider than the polar areas. Thus disks are a bit difficult to prepare. The best recommendation is that you use those supplied by ALPO because they have the properly drawn disk. Also note: Because of the planet's rapid west to east rotation, seen right to left in an astronomical telescope, it is best for the observer using a refractor or SCT to leave out the telescope's star diagonal. This way you will not become confused by the image reversal given by the diagonal.

Begin your drawing (Fig. 9-23) by determining the position of the equator. Quickly fill in the main belts and zones, then add the fainter belts. Before you begin to fill in the detail, note the time and mark it on your form. You will have 10 minutes to place the main details on the belts and zones (Fig. 9-24). You must be constantly aware of the rotation of the planet and keep checking the position of each detail with

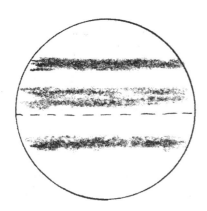

FIGURE 9-23 *Drawing Jupiter I.*

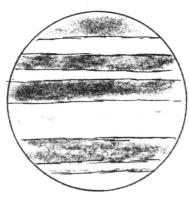

FIGURE 9-24 *Drawing Jupiter II.*

FIGURE 9-25 *Drawing Jupiter III.*

relation to the rest of the drawing. Indicate bright but ill-defined areas and low-contrast features with a dashed line. Use a sharpened eraser to add details to the darker belts. When all the details are drawn, and before your time is up, check the planet again and compare the drawing to the image. Shade in the belts and prominent features (Fig. 9-25), then note the time the drawing was completed. Because of the time limitation imposed by Jupiter's rotation, it takes a bit more practice to get the hang of making full-disk drawings. The only way to become proficient is by constant practice, so don't be disappointed by your early drawings. And by all means save them—no matter how bad you think they look.

Strip sketches are easier to execute. Instead ranging over the entire visible ball of the planet, a strip sketch is simply a continuous record of the details in a particular zone or belt of the planet as it crosses the CM. Time is noted for the transit of each feature in order to compute its longitude, which is labeled across the bottom of the sketch. Strip sketches allow the observer to concentrate on detecting and recording much finer detail than can be seen during a full-disk drawing. Like full disk drawings, strip sketches concentrate on obtaining the position and structure of the object—not producing a work of art.

The technique for making a strip sketch is simple. If you are studying a single zone or belt, the boundaries of the belt are the boundaries of the illustration. If you are studying a complex area such as the SEB with its north (SEBn) and south (SEBs) aspects, indicate which is which by a notation at the edge of its placement on the page. Now sit back and watch the show. As each feature crosses the CM sketch away, moving across the page. Pay close attention so you can obtain as much detail as possible.

Note the time of transit for the center of the feature; if you are totally engrossed, either the preceding or following edge will do. As the feature under observation moves off the CM, another might be ready to take its place. If there is a gap before the next feature reaches the CM, position it on the page a proportional distance from the last feature entered. It's not a bad idea to indicate the transit time of the point midway between the two features to ensure accuracy. The beauty of a strip sketch is that you can extend it over as long a period of time as you wish with no time limit for its completion.

THE SATELLITES OF JUPITER

The show Jupiter puts on for the amateur astronomer is not restricted to its cloudy surface. There is much to see and study in the space around the planet. When Galileo first turned his telescope on Jupiter, he was surprised to find four small stars close by. These four satellites, called the Galilean satellites (Fig. 9-26), are the only ones visible out of a total of 16 plus that circle the planet. Even though they have been studied by the Galileo spacecraft circling Jupiter, they are also the subjects of some basic scientific observations made with the smallest telescope.

In order out from the planet, the Galilean satellites are I, Io; II, Europa; III, Ganymede; and IV, Callisto. During the course of their orbital journeys around the planet, each satellite can be subject to one of four possible events (Fig. 9-27). It can be eclipsed or occulted by the planet, and either the moon itself or its shadow can move across the planet in transit. Each of these events can further be broken down as entry to or exit from the event. These events can be studied with almost any size telescope.

FIGURE 9-26 *Jupiter from the Hubble Space Telescope, showing the small Galilean satellite Io. (NASA)*

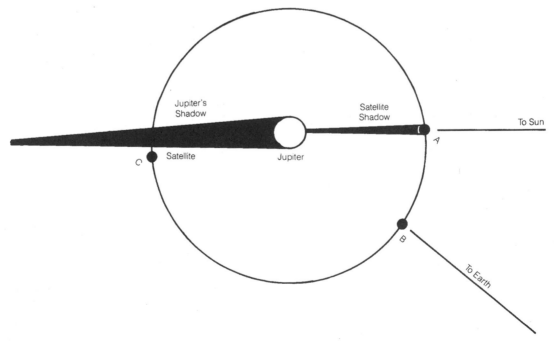

FIGURE 9-27 *The various phenomena of Jupiter's satellites that can be seen from Earth.*

The easiest of the phenomena to observe are the eclipses of the satellites. Since Galileo, astronomers have watched these four satellites moving into and out of eclipses. These movements have given us a chance to study the gravitational effects of one body on another, called *perturbations*, in a closed, rapidly moving system. Observations of these events were also of tremendous value during the seventeenth and eighteenth centuries in refining the positional longitudes on Earth. Today we have more sophisticated ways to determine longitudes on Earth—but the Galilean satellites and their eclipses still provide the best real-time laboratory for determining effects of perturbations.

Since 1975, ALPO has had a program for the timing and analysis of Galilean eclipses. Since the program's beginning, thousands of timings have been analyzed to refine the positions of the four Galileans. Not only is this information of use to theoretical scientists in determining perturbation effects, it is used by professionals. For example, such data was used by NASA to plan and launch the Galileo mission to Jupiter.

The observational requirements for the program are simple. A telescope as small as 50 mm, using at least 60×, and an accurate timepiece are all that are necessary to join in another scientifically useful program. Electronic watches manufactured today can give you accuracy to within one second when they are checked against radio time signals from station WWV. Each year ALPO's Jupiter division, the U.S. Naval Observatory's *Astronomical Almanac*, and the Royal Canadian Astronomy Society's (RCAS's) *Observers Handbook* publish ephemerides for the events of the Galileans for the upcoming year. Smaller, less-detailed listings of the Galileans are often found in such publications as Guy Ottewell's annual *Astronomical Calendar*. Your job is to observe as many of the eclipse events as possible during the next apparition, record the time of the event using your timepiece, and then forward the raw data to ALPO for analysis.

In the field, set up your equipment well in advance of the predicted event (which is listed for UT, so don't forget to make the necessary corrections). You should be ready to go at least 10 minutes before events predicted for Io and Europa, and at least 30 minutes before those for Ganymede and Callisto. Use the extra time to identify the Galilean satellites present around the planet.

Eclipse events that occur before Jupiter reaches opposition cause the satellite to disappear long before it reaches the preceding limb of the planet. For Io and some points in the orbit of Europa, this means that you will probably not be able to see the satellite emerge from the shadow—the event will be blocked by Jupiter itself. The other satellites are far enough away, so they will reappear before passing behind the planet in occultation.

After opposition the reverse is true, with the satellites reappearing well away from the following limb of Jupiter. Disappearances after opposition take place in a small area defined on one side by the limb of the planet and on the other by the shadow of Jupiter. If you are timing a *disappearance*, locate the target moon and watch it fade slowly from view, recording the time to the nearest second when the last speck is visible. When recording the *reappearance* of a satellite from the shadow of Jupiter, take a guess where the satellite will appear and note the time to the nearest second when the first speck of light is visible. Transfer your data to the report form supplied by ALPO and fill in the necessary data on your telescope, seeing conditions, and any information you feel is relevant to your timing. When the sheet is full, mail it to

ALPO and ask for another. They will be only too happy to send it along. Or you can send the information electronically, e-mailing your data to the organization.

Transits of the Galileans and their shadows across the disk of Jupiter are also easy to follow, although your instrument should be a bit larger. While a 60-mm is sufficient to show the entry and exit of the satellite as a bright point of light, a larger instrument will give better resolution when following the satellite completely across the disk. Because the ingress and egress of the satellite require you to resolve its actual disk, an aperture of at least 8 inches is required. The shadow of the satellite will show up as a dark, distinct black spot on the surface of the planet. It is easy to distinguish against the background of belts and zones because it is so black. The satellite itself is more difficult to detect, especially when it is crossing a zone, because the light object can easily blend into the background of the planet.

As with eclipses, the sequence of transit phenomena is dictated by the location of Jupiter in its orbit. Prior to opposition, the shadow precedes the satellite, and its movement onto the disk, or *ingress*, will come before that of the satellite itself. After opposition, the opposite is true—the shadow following the satellite is still seen on the planet after the satellite itself has left the disk, or *egressed*. Timing these events is simple but requires a sharp eye. The crucial points during ingress are when the satellite, or shadow, first touches the disk (first contact) and when the satellite or shadow is seen to be completely on the disk (second contact). Egress is harder to detect because the satellite may blend into the background. Timings should be made when the satellite or shadow reaches the edge of the disk (third contact) and when it is completely clear of the disk (fourth contact). Again, timings should be made to the nearest second.

Occultations of the Galileans are timed in much the same way as those of a star occulted by the Moon. The time is recorded when the satellite disappears behind the planet and again when it reappears on the other side of the planet. The reappearances of the satellites are particularly hard to catch because of the glare from the planet itself.

OBSERVING SATURN

The last planet in the amateur astronomer's solar system is also the loveliest. Saturn, with its ring system, is probably the sight most sought after by people attending public star parties. It is easy to tell when people at the telescope encounter the planet, whether for the first or the umpteenth time. They simply freeze at the eyepiece, not daring to move for fear of losing the wonderful sight. On many occasions, they ask the astronomer if they were really looking at Saturn or a cleverly placed picture of the planet.

Observing Saturn pushes the amateur to the very limit of his or her equipment and abilities. Despite its beauty, Saturn is a very difficult object to observe. When Galileo first turned his telescope on the planet, he was puzzled by its appearance, recording it as a "triple planet." His instrument could not resolve the majestic rings of the planet into anything more than "a large disk, touched on either side by a smaller disk." It would remain for Christian Huygens to solve the riddle of the triple planet some 50 years later.

During its apparitions, Saturn is visible for 10 months out of the year, varying in magnitude from +1.5 to 0. To the naked eye, Saturn is easy to pick out from the background stars as a bright yellow-white "star." The apparent diameter of the disk of Saturn varies from a minimum of about 15 arcseconds to a maximum of 21 arcseconds.

Saturn, like Jupiter, is a gas giant and it shows the characteristic dark belts and lighter zones. But unlike Jupiter, you are going to need a larger instrument to see them. A telescope of at least 4 inches in aperture—and preferably 6 inches or larger—is needed to pick out the subtly colored bands from the bright background clutter. Saturn also does not display the abundant atmospheric activity of its inner neighbor. The most prominent features on the Saturnian disk are white spots and ovals similar to those on Jupiter. But unlike those on Jupiter, these are relatively short-lived, most of them lasting only a few weeks (Fig. 9-28).

Saturn is a planet that almost requires the use of filters during observation. The lack of contrast between the features, when they are there, makes it difficult to pick them out of the background. This is further complicated by the bright image transmitted by the ball of the planet itself. Because the belts on the ball are rather dusky, they are often lost in the glare of the planet. A light blue filter (Wratten #80A) will give you the best chance to separate the belts and zones by increasing their contrast.

ALPO's Saturn Section distributes a complete kit of materials, including planetary blanks for all the aspects of the rings, and information on the planet's elusive features. If you want to challenge your observing skills, join ALPO and get into the Saturn Section.

THE RINGS OF SATURN

The main focus for a small telescope pointed at Saturn is the rings (Fig. 9-29). Less than 100 feet thick, they measure 175,000 miles across and represent a celestial hail

FIGURE 9-28 *A storm, similar to a thunderhead on Earth, on Saturn is caught with the Hubble Space Telescope in December, 1994. (NASA)*

storm whipping around the planet at 90,000 miles per hour. The brightness of the planet, as we see it with the naked eye, is due in large part to the reflective properties of the rings. Saturn takes 29.5 years to circle the Sun. Twice during that time, once after a 13.75-year period and again after a 15.75-year period, the Earth's view passes briefly through the plane of the rings, and we see the rings edge-on. For this period of time, Saturn loses its most prominent feature, and the ball of the planet is bare in the telescope. Prior to this event, the rings appear to slowly close and there is a resulting loss of brightness when viewed with the naked eye. But the same orbital mechanics that briefly take the rings away also show them to us in all their glory. At a point between each plane passage, the planet is tipped toward Earth by 27 degrees, displaying almost the entire ring system. During the 13.75-year period the ring system's southern face is presented to Earth, while during the 15.75-year period the northern face of the rings are seen.

If you look at the planet before opposition, you can detect the shadow of the planet as a jet black notch on the western side of the rings. After opposition, the notch moves around to the east side. Associated with the shadow is a bright white spot, called the *Terby White Spot*. This spot is thought to be caused by electrical activity brought on by the shadow's passage over the rings.

FIGURE 9-29 *The many ringlets in the rings of Saturn, taken by the Voyager spacecraft. (NASA)*

The rings of Saturn are not a single feature, but a system of rings and gaps that circle the planet. From Earth the amateur can easily detect the three main rings, labeled A, B, and C. The outermost ring, A, is divided into two sections: the dark outer one and bright inner one. The bright *annulus* is separated by a fine, dark line called *Encke's division*. Theoretically a difficult object to see with an 8-inch telescope, Encke's division has been reported by numerous observers using 6-inch instruments. These reports are confirmations of the dictum stated by Sir John Herschel, William's son, that "when an object is discovered by a superior power, an inferior one will suffice to recover it."

Separating rings A and B is a prominent dark division called *Cassini's division*. The brighter inner surface of ring A, as seen by a 6-inch telescope, exhibits an irregular appearance along the edge fronting Cassini's division. Cassini's division itself is easily seen in a small telescope as a dark gray gap between the two rings.

Ring B is the brightest and largest of the three major rings. Its appearance in a telescope shows its brightest at its outer edge, then darkening as it nears the planet. As many as seven dark fines can be seen in Ring B with telescopes as small as 5 inches in aperture.

The innermost ring, Ring C, is also called the *Crepe ring*. Having little contrast against the dark background of the space between the planet and the rings, it is most easily seen as it crosses the disk against Saturn's bright background.

THE SATELLITES OF SATURN

Saturn is blessed with over 20 satellites (thought to be the most in the solar system), 5 of which can be seen with a small telescope. The brightest of these, Titan, is the largest satellite in the solar system; it appears as a magnitude-8 star orbiting the planet in 16 days. Orbiting closer to Saturn and in 5 days is the moon Rhea, visible as a magnitude-10 "star." Inside the orbit of Rhea, an amateur with a 4- to 6-inch telescope can locate the moons Tethys and Dione. Tethys, orbiting the planet in 2 days, appears as a magnitude-10 star—as does Dione, which completes a circuit of Saturn every 3 days.

The outermost visible moon, Iapetus, requires almost 80 days to orbit the planet and has a variable magnitude ranging from 10.1 to 11.9. The variability of Iapetus is caused by differences on the surface of the satellite's two hemispheres. One is covered with highly reflective ice, while the other side has been stripped of this ice cover possibly by some type of reaction between the Sun's rays and the surface.

SIGHTING ASIDES

You may find another small group of astronomers who want to hunt for other celestial objects. For example, there are many amateurs who keep track of Uranus and Neptune; try contacting ALPO for more information of remote planets. (Fig. 9-30) Sometimes amateurs will watch for artificial satellites in the nighttime sky—including times when the space shuttle may go overhead. Look and ask around. There are sites on the Internet, not to mention some people in astronomy clubs, who are dedicated to watching these dimmer objects of the nighttime skies.

OBSERVING FORM
A.L.P.O. Remote Planets Section

General Information

Name _____ Location _____

Date (U.T.) _____ Start _____ U.T.   Finish _____ U.T.

Telescope: Type _____ Aperature _____ Magnification _____

Seeing _____ Transparency _____ Your Latitude _____

Type of Observation (circle) A, B, C, D, E-1, E-2 (see below)

A Visual Observations

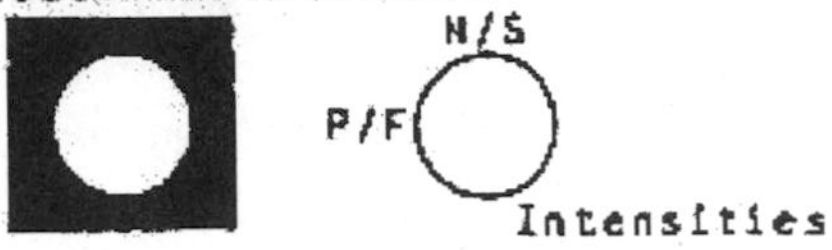

N/S

P/F

Intensities

Planet_____

Circle N or S and P or F

Comments: _____
_____
_____
_____
_____
_____
_____

C Color Estimate

Planet_____

Color description: _____
_____

B Photography/CCD image

Method: (please circle)

prime focus/eyepiece proj/CCD

film _____

Exposure time _____

f ratio _____

Developer_____

Comments_____

D Occultations/Near misses

Planet_____

Star occulted _____

Planet: RA_____ Dec _____

Star: RA_____ Dec _____

Comments_____
_____
_____

E-1 Photoelectric Photometry (Reductions done on separate sheet)

Time UT	Star/Planet	Filter U B V R I	Scale	Integration Time	Count	Sky brightness

E-2 Visual Photometry

Comparison Star 1 (HD or SAO #)_____ Mag.* _____ Source _____

Comparison Star 2 (HD or SAO #)_____ Mag.* _____ Source _____

Planet _____ Estimated Mag.* _____ (Mag.=Magnitude)*

FIGURE 9-30 *The ALPO chart for remote planets, such as Uranus and Neptune. (NASA)*

10

Observing Comets, Asteroids, and Meteors

The heavens themselves blaze forth the death of princes.

SHAKESPEARE

Most professional astronomers regard comets, asteroids, and meteors as annoying little smudges and streaks of light that intrude on their photographic plates. But there has always been a special bond between amateur astronomers and the "rubble" of the solar system.

While most professional astronomers tend to ignore these small members of the Sun's family, amateurs seek them out. An amateur will spend countless hours searching for a new comet or the passing of a near-Earth asteroid, or an entire night counting meteor trails emanating from a meteor shower. Each offers a different reward—and each offers the amateur a way to contribute to science.

COMETS

Amateur astronomers and the public react differently to comets. To most people, Comet is a household cleaner. And until recently, with the comets Hyakutake and Hale-Bopp, few members of the nonastronomical public have seen a real live comet. They may have seen pictures of one, usually comet Halley in 1910, but pictures don't count.

People respond with interest to cometary announcements because they want to see the spectacle presented by a really bright object. On the appointed day they dutifully go outside and look toward the appropriate place in the sky. But after Hale-Bopp was viewed for months, the public still went back inside. If only all comets

169

were bright and clear like the comets Hyakutake and Hale-Bopp—the public would be out watching the sky every night. But these two comets were exceptional.

Amateurs don't react that way. They are out there looking not only at the spectacular comets, but also for other comets, no matter how faint. They look because there is no one else out there to look and because each comet gives us a chance to peek at the history of the solar system. Professionals may want to turn their instruments on the comet as it streaks toward the Sun, but the time on their telescopes is too expensive to divert them to watch for very long. Exceptions do exist, including the return of comet Halley in 1986 (Fig. 10-1); comet Hyakutake in 1996 (Fig. 10-2); and comet Hale-Bopp in 1997 (Fig. 10-3). But after the comets streamed by, interest dropped in the public sector. And again, the comet watchers, mostly amateur astronomers, are the only ones out looking for the next bright or dim comets.

COMETARY GEOGRAPHY

The origin of most comets is thought to be far away, in the distant cold of the *Oort cloud*, a theoretical huge shell of material left over from the formation of the solar system. The cloud is a lumpy mass of ice and dirt orbiting the Sun, with the comets

FIGURE 10-1 *Comet Halley, taken on March 8, 1986, on Easter Island as part of the International Halley Watch Large Scale Phenomena Network. (W. Liller and NASA)*

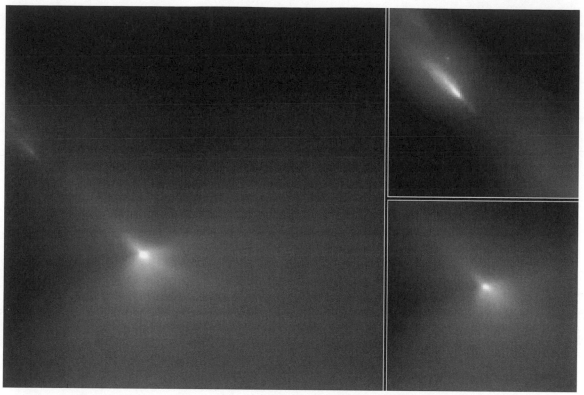

FIGURE 10-2 *Comet Hyakutake through the Hubble Space Telescope when the comet passed at a distance of 9.3 million miles from Earth. (NASA)*

making up the shell too small and faint to see from Earth. But occasionally, something happens, such as a faint tug or a push by a more massive object, and a comet finds its orbit changed ever so slightly. The long journey toward the Sun has begun.

On its way to the nearer reaches of the solar system, the heart or nucleus of the comet begins to change. The driving rain of radiation from the rapidly brightening Sun causes it to heat up. Gases begin to bubble and boil off. A cloud of gas called a *coma* surrounds the nucleus and reflects the light of the still distant Sun, making the comet visible against the darkness of space. From Earth, its light may be detected on photographic plates made at an observatory or it may appear as a faint smudge in the field of an amateur astronomer's telescope. Regardless of who sees it first, a new comet is discovered, confirmed by other sightings, and given a name.

During the course of their brief and rapid race around the Sun, comets carry three names. The first is the designation of the year of discovery and the order of its appearance after other comets discovered that year. For example, the first comet discovered in 1999 would carry the designation 1999a, the second 1999b, and so on. The comet's second name is the name of its discoverer. For example, the tenth comet discovered in 1988 was found by California amateur astronomer Don Machholz. This comet now carries the names 1988j and Comet Machholz. When it reached its closest point to the Sun, or perihelion, Comet Machholz received its third name—the year it

FIGURE 10-3 *Comet Hale-Bopp in April 1997. (Courtesy of Alan Hale, Southwest Institute for Space Research)*

reached perihelion followed by a roman numeral indicating its order among all the comets reaching perihelion that year.

As it approaches the Sun, the comet changes. The coma and the nucleus it hides are now called the *head*. The comet is reacting to the incoming radiation from the Sun, causing more of the nucleus to boil off, carrying dust and dirt away from the head. This dust is caught by the *solar wind*, highly energized particles thrown out from the Sun, and is pushed away from the comet's head, where it begins to glow from the radiation. We see this glowing dust as the *dust tail*, which points in the direction of the solar wind, away from the Sun.

A comet really has more than one tail. In addition to the dust tail, a gas or *ion tail* can be seen streaming out directly behind the comet's head. Sometimes another tail, an *anti-tail*, is seen pointing toward the Sun. The anti-tail is nothing more than an optical illusion created when we look through the plane of the comet's orbit at the dust tail. Comet Hale-Bopp was known for its multiple tails.

As the comet sweeps around the Sun, it may or may not reach the safety of the Oort Cloud again. Minute tugs of gravity from each planetary body it passes during its journey toward the Sun act to change its orbit. This orbit determines the ultimate fate of a comet (Fig. 10-4). If the orbit is an elongated circle, called an *ellipse*, the comet will swing around the Sun and head back toward the Oort Cloud. It may not make it that far, however, because its speed has been affected by the huge gravitational pull of

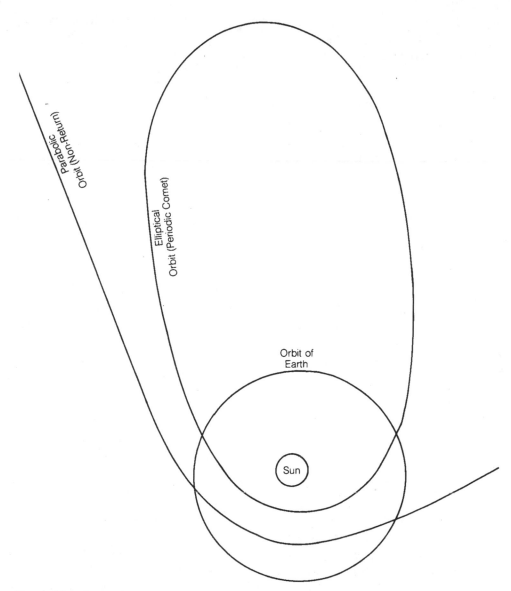

FIGURE 10-4 *Comet orbits and the Oort cloud.*

the Sun. As it heads away from the Sun it begins to slow down. At some point in its orbit it will stop and begin to fall back toward the Sun. A comet with this type of orbit will return to the Sun again and again until it ultimately melts away to almost nothing.

These comets are called *periodical comets* because they return at definite periods. Comet Halley is a periodic comet. Its orbit takes it out toward Neptune over a period of 76 years. Comet Encke is another periodic comet, but it doesn't venture as far from the Sun. It only goes out as far as the orbit of Jupiter and has a period of 3.3

years. The names of all periodic comets are preceded by the designation "P/," so comet Halley becomes P/Halley.

Periodic comets fall into two classes. *Short-period comets*, such as Halley and Encke, have periods of 200 years or less; comets with periods of over 200 years are called *long-period comets*. Comet Hale-Bopp is a long-period comet, with an orbit of about 4,200 years—so it's been in this solar system before.

THE ROLE OF THE AMATEUR

Amateurs have always had a special relationship with comets. Not only do they provide valuable information on the structure of comets, amateurs very often make the ultimate contribution, the discovery of comets. Comet hunting has always been an area in which amateurs have excelled. Many comets are discovered by accident when they turn up as smudges on photographic plates made by professional observatories. Satellites such as the infrared astronomical satellite (IRAS) have even been the first to detect comets on their way toward the Sun, but these discoveries also have been accidental. Amateurs throughout the world, however, are at their telescopes every clear night with the purpose of discovering comets. One of the best-known amateur astronomers is David Levy of Tucson, Arizona, who has logged thousands of telescope hours hunting comets. His efforts have been rewarded with many comets that carry his name—including comet Shoemaker-Levy 9, the fragmented comet that crashed into the atmosphere of Jupiter in 1994 (Fig. 10-5).

Comet hunting is not for every amateur. If you want to hunt comets to gain fame and fortune, you are going to be disappointed. The hours are long, and there is no guarantee that you will ever find a comet. Don Machholz searched for 1,700 hours before he found his first comet. David Levy spent over 900 hours at his telescopes before he bagged his first comet. These numbers, when combined with such mundane things as working a regular job and having a social life, tend to discourage all but the most dedicated observers.

If you insist on going out tonight and looking for a comet that will carry your name for all eternity, there are a few things you can do to help the process along. Comet hunting is a competition between amateurs and professionals. While most professional astronomers discover comets during the pursuit of other projects, a few have formed organized patrols and regularly use their larger, more sensitive photographic instruments in an attempt to recover periodic comets headed back toward the Sun.

FIGURE 10-5 *The comet Shoemaker-Levy 9 broke into 21 fragments in May, 1994, seen in this Hubble Space Telescope image; the broken comet impacted Jupiter in mid-July, 1994. (NASA)*

They are looking to confirm the computations made for the orbits of these comets. If they discover a new comet in the process, it's "icing on the cake," so to speak.

Most of these professional efforts are centered on areas surrounding the meridian, where the photographs can record objects as faint as magnitude 20. It would seem obvious that if you search this area, you are searching against some heavy opposition. The most fertile area of the sky, however, is the area that professional photographic patrols have to avoid. Recording objects within 45 degrees of either side of the Sun reduces the effectiveness of these professional instruments. Exposures made with the Sun just below the horizon can cut the limiting magnitude of the photograph and possibly damage the expensive, sensitive instruments. So the area near the Sun, just after sunset or before sunrise, is best for visual searching.

The tools of the comet hunter may include any sort of optical aid. You can even use the naked eye to search for bright comets near the Sun. William Bradfield of Australia, discovered most of his comets using everything from 7×35 binoculars, with which he discovered a comet in 1980, to a 10-inch reflector, with which he discovered a comet in 1984. Leslie Peltier discovered 12 comets using a 6-inch f/8 refractor as his prime instrument. David Levy used a 16-inch f/5 reflector as his primary instrument.

In comet hunting, the altazimuth mounting is a definite advantage over the equatorial because you will be observing near the horizon. The altazimuth allows you to easily scan along it without putting a strain on your back or your patience.

Focal ratio, although not crucial, is important. It must be remembered that you are going to be looking for a very diffuse, nebulous object. While the images provided by f/4 and f/10 systems will look equally bright (provided the magnification is the same), the f/4 will provide a wider area of the sky in your field of view and will make the comet stand out from the background. The f/10 covers a smaller area and, because it provides an appreciably darker field with a corresponding decrease in the brightness of an object, makes a suspected comet blend into the background.

It is best to use an eyepiece that gives a wide, flat field and a magnification of no more than $2D$. Erfles are hard to beat in this category, although designs like the Tele-Vue wide-field and the Nagler-2 and -4 types are also useful. If you are using a short-focal-length newtonian such as a 6-inch f/4, you will want an eyepiece with a focal length between 40 and 48 mm. You can also use any telescope or pair of binoculars to find a comet. The important thing is how you use the instrument.

WHEN AND WHERE TO LOOK

The best time to look for comets is either just after sunset or just before sunrise. Of the two, the time before sunrise yields the most new comets. A problem you will have to contend with is the Moon. When the Moon is beginning a new lunation and is rising brighter each night, it is time to search the sky before sunrise. As the Moon starts to spill over to the morning sky, try to time your searches to begin with moonset.

For the period just before a new Moon, when it rises about 2 hours before the Sun, begin your search about 3 hours before sunrise. This will give you at least an hour to search. If you are searching the sky after sunset, the best time to look begins about 2 days after a full Moon. This gives you at least an hour and a half from sunset until moonrise to hunt. As the Moon rises progressively later each night, you have more time each night for your search.

The area of your search should extend from the horizon to a point about 45 degrees above the horizon and within 45 degrees on either side of the point where the Moon has set or will rise. Cover the area with overlapping sweeps, each sweep arc containing slightly less than the field covered by your telescope (Fig. 10-6). For searches after sunset begin at the horizon and work up. Morning sweeps should begin at a point about 45 degrees above the horizon and work down.

What you are looking for is a little fuzzy patch of light (Fig. 10-7). During the course of your sweeping you will come across dozens, even hundreds of objects that fit that description. The skies are filled with "little fuzzy patches," seemingly put there to confuse and confound the comet hunter. Charles Messier, the French comet hunter of the late eighteenth century, became so frustrated with these little fuzzies that he began to list them "so that astronomers would not confuse these nebulae with comets just beginning to shine." Today we know this list as the *Messier Catalogue* (listed in Appendix E). It includes 110 of the finest deep-sky objects in the heavens. To better prepare yourself and learn what comets do not look like, check out the members of this list before you start hunting.

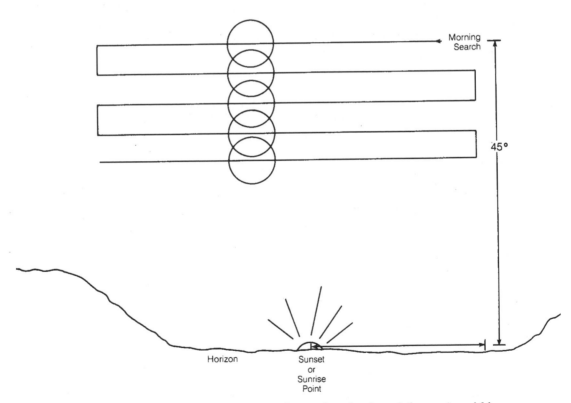

FIGURE 10-6 *Search pattern for comets. For a morning search, start about three hours before sunrise and follow the pattern. For an evening search, begin about one hour after sunset and star at the horizon working up to 45 degrees. The fields of view for each sweep should overlap.*

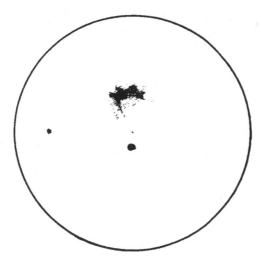

FIGURE 10-7 *Sketch of a comet as it appears in an amateur telescope.*

In addition to the 110 objects on Messier's list, another 8,000 galaxies, nebulae, and star clusters are listed in the *New General Catalogue of Nonstellar Astronomical Objects*, called the NGC for short. In addition to the NGC there are the *Uppsala General Catalogue of Galaxies* (UGC), the *Catalogue of Selected Non-UGC Galaxies* (UGCA), the *Catalogue of Galactic Planetary Nebulae* (PK), and so on. You get the idea.

There are hundreds of objects out there you might confuse with a comet. Some areas are so choked with faint galaxies, such as Coma Bernices, Leo, and Virgo, that it doesn't pay for the comet hunter to search there. So what do you do to sort everything out?

Get yourself a good star atlas. An atlas like Wil Tirion and Roger W. Sinnott's *Sky Atlas 2000* is an indispensable asset for the comet hunter. Not only does it list stars down to magnitude 8, it also lists thousands of objects in the NGC and other objects. *Uranometria* lists stars down to magnitude 9.5 and thousands more deep-sky objects than *Sky Atlas*. A good atlas gives the comet hunter a reference to check quickly if he or she finds something suspicious. In many cases, amateur comet hunters prefer photographic atlases over drawn or computer-generated atlases—both of which can have glaring omissions. And if what you see is listed, it's on to the next field. If it's not, . . .

REPORTING YOUR FIND

Stay calm and check the image of your "comet" carefully. Is the object real or a light reflecting off a nearby star—sometimes referred to as ghost images? Move the object around the field and see what happens. Check the object with a higher-power eyepiece. Is it a faint star near the limiting magnitude of your telescope or a group of stars? Both types of objects can look diffuse and nebulous. Draw a chart showing the object and the surrounding background stars. Be careful to get the distance relationships between the stars and your object correct. You can then transfer this location onto your star atlas and get an accurate right ascension and declination from it.

Comets move very quickly. By checking the drawn field with the field of view just 15 minutes later, you might be able to detect movement. If you are not sure, wait. Movement and the direction of that movement will be important when you finally report your find. So make sure it is moving and that you plot the direction of movement on your chart. Does it have a tail? The tail may be faint, but if it is there along with movement—that's usually the clincher.

Estimate the magnitude of your comet by comparing it with other stars in the field that you have already identified from your atlas and chart. The best way to do this is to memorize how the comet looks and throw the field slightly out of focus. Compare the image of the focused comet to the now diffuse star images. Then decide which star is closer to the magnitude of the comet.

You're almost ready to take the next step, but don't count your comets before you're absolutely sure. Yes, it might be a comet, but it might not be yours. It could be a previously listed comet or a periodic comet on its way around the Sun. And remember, at any given time, there are usually at least two or three (often more) comets visible in the night sky easily within the visual range of an 8-inch reflector.

Check sources such as the *International Comet Quarterly*, which lists orbital elements and ephemerides for all periodic comets due during the year. Also check the most recent *IAU Circulars*—to which you can subscribe and which your can receive either via mail or electronic mail through the Central Bureau of Astronomical Telegrams (CBAT) in Cambridge, Massachusetts, a clearinghouse for discoveries under the International Astronomical Union (IAU). These circulars announce many astronomical discoveries (comets and other objects) and send them to subscribers complete with ephemerides. The CBAT also publishes the *Catalogue of Cometary Orbits*.

Here is a sample of an *IAU Circular*:

Circular No. 6123

Central Bureau for Astronomical Telegrams
INTERNATIONAL ASTRONOMICAL UNION
Postal Address: Central Bureau for Astronomical Telegrams
Smithsonian Astrophysical Observatory, Cambridge, MA 02138, U.S.A.
IAUSUBS@CFA.HARVARD.EDU or FAX 617-495-7231 (subscriptions)
Phone 617-495-7244/7440/7444 TWX 710-320-6842 ASTROGRAM CAM
MARSDEN@CFA.HARVARD.EDU or GREEN@CFA.HARVARD.EDU (science)

SUPERNOVA 1994ak IN NGC 2782

M. W. Richmond, Princeton University, and R. R. Treffers, S. D. Van Dyk, and A. V. Filippenko, University of California at Berkeley, confirm that there is a new starlike object in NGC 2782, presumably a supernova, close to the position reported on IAUC 6122.

Observations were conducted as part of the Leuschner Observatory Supernova Search, which uses an automated 0.76-m telescope equipped with the Lawrence Berkeley Laboratory CCD camera. The supernova (R = 17.4 ± 0.8) is located about 58″ west and 33″ south of the galaxy's nucleus in an image obtained on 1994 Dec. 30 UT; it is also barely visible in a mediocre image obtained on Dec. 23, but not visible (upper limit R = 19) in a good image on Dec. 1.

S. Nakano, Sumoto, Japan, forwards the following additional unfiltered CCD magnitudes of SN 1994ak obtained with a 0.25-m Schmidt-Cassegrain by Y. Kushida at Yatsugatake South Base Observatory: 1994 Dec. 25.780 UT, 15.9; 28.591, 16.0; 29.619, 16.0; 31.616, 16.1; 1995 Jan. 1.645, 16.3; 2.600, 16.3.

D. Balam, Climenhaga Observatory, Victoria, provides the following magnitude and precise 2000.0 position for SN 1994ak obtained on Jan. 3.44 UT: V = 18, R.A. = 9h14m01s.47, Decl. = +40∞<Judy: lower case o on hard copy but underline char. in file. Took a guess at a degree sign.>06'21".5; the supernova is indicated as being about 43" west and 27" south of the galaxy's center.

SUPERNOVA 1994ai IN NGC 908

M. Cavagna and E. Galliani, Sormano, Italy, report the following precise position for SN 1994ai, obtained on Jan. 1.80 UT: R.A. = 2h23m06s.17, Decl. = -21∞13'58".3 (equinox 2000.0).

V1251 CYGNI

This cataclysmic variable (cf. IAUC 5377) is evidently in outburst, as indicated by the following visual magnitude estimates: 1994 Dec. 22.736 UT, [13.6 (P. Schmeer, Bischmisheim, Germany); 30.756, 13.3: (Schmeer); 31.713, 13.4 (L. Szentasko, Veresegyhaz, Hungary); 31.833, 13.2 (Schmeer); 1995 Jan. 1.771, 13.7 (Schmeer).

The following precise position for V1251 Cyg has been measured by B. Manning, Stakenbridge, England: R.A. = 21h40m54s.36, Decl. = +48∞39'43".1 (equinox 2000.0).

1995 January 4 (6123) Daniel W. E. Green

At this point, you may also want to contact (privately) a local observatory or some other amateur astronomers just to confirm your discovery. But it's best not to put the information over the Internet—such a sudden surge in people hunting for the object may lead to wild goose chases or the chance that you will lose credit for the discovery. And remember, the only way to make sure the discovery will be yours is to contact the CBAT first.

If the object you are viewing is moving (with or without a tail), is not a listed or misidentified deep-sky object, and is nobody else's comet—it could be all yours. High-tail it into the house and call Western Union or get on the Internet (see the e-mail address at the back of this book)—it's time to inform someone. If you are absolutely sure, it may be time to call the CBAT.

If you are absolutely sure that you are seeing a new comet, the best and fastest way to inform the CBAT is via the Internet. They offer a discovery form or you can e-mail the people in charge of the CBAT. The preliminary information you must include is as follows:

- Your name
- Your address and contact details (preferably e-mail address, otherwise telephone/fax number)

- The date and UT time of observation (use Universal Time, not local time)
- The observation method (e.g., naked eye, visual telescopic observation or telescopic CCD)
- Specific details on instrumentation (size, focal ratio, etc.) and exposures (type of film or CCD, length of exposure, etc.)
- Observation site

You'll also need to know the position of the comet—accurate to at least 1′ in declination and 0.1 minute of time in right ascension. This is not difficult, especially with all the good atlases and grid overlays available—speaking of which, you will want to give a list of which sources you checked to confirm the comet. If you have a photograph or especially a CCD, you should record accurate positions to 0.01 second of time for the right ascension and to 0″.1 for the declination. You can also include some details of the object, such as the magnitude, colors, possible size and shape, amount of diffuseness, if there is any tail (and the types of tails), and any other preliminary observations.

Many amateurs usually wait another night before reporting the comet. That way, they can confirm the discovery. Or, if you do search and discover only one night, observe the comet with many different instruments if possible—and try taking multiple photos or CCD images of the object. The whole idea is to confirm and verify your discovery as much as possible. It will save you a great deal of grief—and save time for larger observatories that search for your comet. No one wants to find out that what he or she "discovered" was a merely a globular cluster.

TRACKING A COMET

Once a comet is discovered, what can the amateur do to continue the study? The most important role the amateur can play is to keep a continuous watch on the comet with an eye toward its development and changes. Detailed drawings of a comet can show the growth or decline of the tail and the corresponding changes in the head of the comet.

The Comet section of ALPO recommends that observers watch for several details (Fig. 10-8). The size of the coma should be estimated or measured during each observation session. This can be done by comparing the coma size to the distance between two stars. You can also use a micrometer reticle eyepiece (see Chapter 11), which allows you to align the coma across a graduated line for a more precise measurement. You can measure the changes in brightness by estimating the coma's *degree of condensation* (DC) on a scale from 0 to 9, with 0 being a diffuse, uniformly bright coma and 9 being a sharply defined, starlike coma.

Make an estimate of the length of the tail in degrees and tenths of degrees. To obtain this information, accurately plot the comet on a star chart and then measure the desired dimensions directly from the chart. Also, try to determine the *position angle* (PA) of the tail. To do so, measure eastward along the coma from the north. Use 90 degrees as east, 180 degrees as south, and 270 degrees as west. If you can't figure out which way is north, give the telescope a nudge toward Polaris. The edge of the field through which stars enter is north. Check the tail to see if it is smooth or if it has developed any condensations or twists since your last look.

| COMET VISUAL OBSERVATION REPORT FORM |

Send copies of completed forms to: INTERNATIONAL COMET QUARTERLY
    A.L.P.O. COMETS SECTION   c/o Dan Green
    Don Machholz, Recorder   Smithsonian Astrophysical Observatory
    P.O. Box 1716   60 Garden Street
    Colfax, CA 95713   Cambridge, MA 02138

PLEASE print or type! Use one Form for each comet. Record Dates to 0.01 day.
Magnitudes to 0.1 mag. Coma Diameters to 0.01', and Tail Lengths to 0.01°.
Include drawings and/or additional remarks on separate sheets of paper.
Site Locations and Sky naked-eye Limiting Magnitudes are not required by ICQ.

COMET _____ 19 ____

OBSERVER: _____ ADDRESS: _____

ICQ ID: _____ Observations in 19__. _____

DATE (UT)		MAGNITUDE			INSTRUMENT				COMA		TAIL		SKY	ST	REM
Month	Date	M/M	Mv	Ref	Apr(cm)	Type	F/	Pwr	Dia(')	DC	Len(°)	PA(°)	L/M	#	#
	·		·						·		·				
	·		·						·		·				
	·		·						·		·				
	·		·						·		·				
	·		·						·		·				
	·		·						·		·				
	·		·						·		·				
	·		·						·		·				
	·		·						·		·				
	·		·						·		·				
	·		·						·		·				
	·		·						·		·				

SITE # LOCATIONS and REMARKS	

FIGURE 10-8 *A comet report form from ALPO. (Courtesy of ALPO)*

Can you search for comets? No doubt you can—and join the ranks of many other amateurs. The most prolific amateur comet hunters today include David Levy (with 8 comets and counting, not including those he found with Eugene and Carolyn Shoemaker); Don Machholz (with 10 to date, and called North American's leading amateur visual comet discoverer); Australian amateur William Bradfield (with 17 to date, and called the world's foremost living amateur visual comet discoverer); Howard Brewington, with 4 comets since 1989—with more no doubt on the way; and Yuji Hyakutake of Kagoshima, Japan, who discovered 2 comets only five weeks apart in 1995 and 1996 (his comet C/1996 B2 was one of the brightest naked-eye comets of the last half century). What are the main ingredients in all these discoveries? Time, patience, and perseverance.

ASTEROIDS

History of Minor Planets

The asteroids are a relatively new addition to the astronomical realm, at least in terms of humans noticing the small bodies. The first and brightest, Ceres, was found by Italian monk Giuseppe Piazzi on the night of the first day of the new year: January 1, 1801—about 200 years after the advent of the telescope.

Most of the asteroids revolve around the Sun in a wide band (the asteroid belt) between the orbits of Mars and Jupiter (Fig. 10-9). Occasionally, some will stray, pulled

Figure 10-9 *The main belt asteroid Ida. (Courtesy Peter Thomas, Cornell University)*

by other astronomical forces—including the planet Jupiter or collisions within the asteroid belt—which force the asteroid into an eccentric orbit. Some of these, in turn, end up as Mar-crossing, Jupiter-crossing, or whatever planet-crossing asteroids. The Earth is not exempt: There are also thought to be at least 2,000 or more near-Earth asteroids crossing or coming close to the Earth's orbit; at this time, only about 10 percent have been discovered. And we also know that such asteroids have struck our planet in the past: We have at least 140 known impact craters on the Earth (Fig. 10-10).

The origin of the asteroids is still somewhat debated. Almost everyone agrees that these rocks are remnants of a planet. But it is not known whether the planet just didn't form as all the other planets coalesced or broke apart. What we do know is that asteroids have been around since the beginning of the solar system, more than 4.5 billion years ago. This is also why scientists are interested in obtaining a chunk of an asteroid—literally, a fossil of the past, held pristine in orbit around the Sun for billions of years.

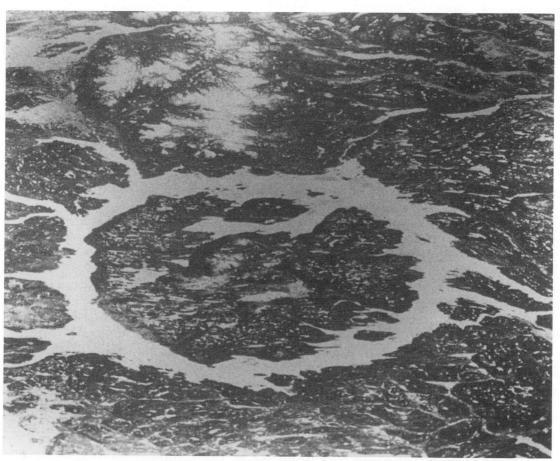

FIGURE 10-10 *The Manicouagan impact crater in Quebec, Canada, is one of the largest on Earth. (NASA)*

THE AMATEUR ASTEROID HUNTER

One reason most amateurs stray away from hunting for asteroids is magnitude: Few asteroids are ever brighter than eighth magnitude, even if they are in opposition. Another reason is that asteroids only show up in the sky as a starlike object; only Ceres, the largest, and one or two others can be seen as disks under good viewing—even with the largest telescopes. On the brighter side, because most are found along the ecliptic, you don't have to search the entire sky to spot an asteroid. And it's always a challenge to add another asteroid to your astronomical "life list."

To identify asteroids, you can try the usual method of consulting an ephemeris, noting the position of the asteroid among the stars, and searching for the elusive body. You may even be lucky enough to spot the yellowish cast many asteroids seem to give. But in most cases, it's difficult to make a positive identification.

There are two ways to find an asteroid (a third one would include photographing part of the sky and finding the tell-tail streak of the asteroid on your film, but we won't go into that in this text). With the first, you compare the telescopic view with the star chart, identifying the wayward "star" in your field. Another method takes more time, but also works: watching a "star" in your field "wander" across the sky.

Using a star chart, you will need to obtain an astronomical chart with stars at least down to magnitude 10—mainly because few asteroids are ever brighter than magnitude 8. Even then, you will probably need a good ephemeris, using the data to plot the area where the asteroid is located on your star chart. From there, you can often identify the minor planet by its brightness compared to surrounding stars. But if you are looking at an area of stellar brilliance, such as in the direction of Sagittarius, the task becomes more difficult.

Thus you may want to turn to the identification of an asteroid by orbital drift—or reobserving a star to see how it's progressed "past" the background of stars. The best time to observe this is when the asteroid is near opposition, when the body moves across the celestial sphere at about ¼ degree per day (about half the Moon's diameter). If it is moving this fast, relatively speaking, you should see some sort of movement if you observe 2 hours after your first viewing. In that time, it will have moved about 1′ of arc relative to a surrounding star.

Some amateurs draw the area they suspect is the asteroid; then, 2 hours later, go back and compare the difference in the drawing with reality. But this is often difficult—sometimes it appears that all the stars are different when viewed 2 hours later. That is when the transit method comes in handy: in a low-powered eyepiece, put in a single hair in a north-south direction; then another hair in a east-west direction (cross-hairs). That way, the image is divided into four quadrants—a much easier way to draw and track a suspected asteroid.

Although asteroid hunting is a challenge for amateurs, some people do just fine seeking the smaller bodies. For example, Arizona amateur astronomer Roy Tucker became the first to discover an 18th-magnitude, Earth-orbit-crossing minor planet June 28, 1997, using "amateur astronomy" equipment. This included a Celestron 14-inch telescope and a CCD camera of his own design. Thus Tucker became the first recipient of the James Benson Prize—awarded to the first 10 amateur astronomers to locate Earth-crossing minor planets (Fig. 10-11).

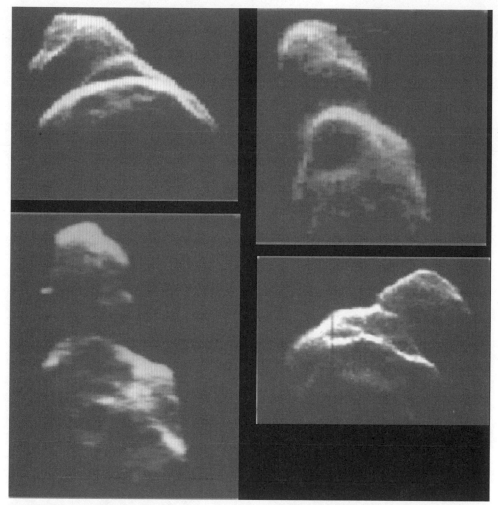

FIGURE 10-11 *Earth-based radar at the Goldstone Deep Space Network took this image of the asteroid 4179 Toutatis in 1992. (NASA)*

METEORS

What if you have no telescope or binoculars, but want to make a contribution, any kind of contribution, to astronomy? Well don't fret. Meteor observing fits right into that category.

Meteors are streaks of light flashing across the sky. They are small particles of dust and dirt that run into the Earth's atmosphere, burning up from the friction. Meteors can appear in the sky as random events called *sporadic meteors*, or they can come from a common point in the sky, called a *radiant*, and be associated with a *meteor shower*. Showers are usually the leftover debris of a comet that has disintegrated after too many trips around the Sun. Remnants of the comet, in the form of dust and minute solid particles, remain in the orbit once occupied by the comet. If the Earth happens to travel through the plane of this orbit, we see a meteor shower—which is why most meteor showers are readily predictable. And if a meteor makes it through the atmosphere, it's called a meteorite (Fig. 10-12).

The equipment needed for meteor observing is simple. All you need is a reclining lawn chair, a clipboard, paper, a pencil, and an optional star chart. The procedure for observing is just as simple: Select an area of the sky, sit back, and make a note of each meteor crossing the area under observation. Despite its apparent ease, meteor observing, especially during times of a shower, can provide useful information. The procedure is the same, but there are a few more things to do.

First, try to determine if the meteor is a member of a shower or sporadic. To do so, hold your clipboard, or any straight edge, along the path of the meteor and note where it came from. Shower meteors will appear to originate from a radiant in the same general area of the sky. The shower is named for the constellation that contains that radiant. The Perseids, a meteor shower that occurs during August, all originate from a spot in the constellation Perseus. Table 10-1 contains a list of the major meteor showers taking place each year. Any meteor you trace back to that constellation using your straight edge will be a member of that shower. Mark any shower meteor

FIGURE 10-12 *This meteorite has caused a stir: Not only is ALH84001 thought to be from Mars, but it may contain evidence of life—a highly debated idea. (NASA)*

with an abbreviation denoting the shower to which it belongs, such as PE for Perseid; mark any sporadic meteors with the letter "S."

You should also estimate the magnitude of each meteor you see. As the meteor crosses the sky, it will appear to grow brighter before it fades out. Estimate its magnitude at the brightest point of the meteor's path. This is simply a matter of comparing that brightness to stars in the area. If a meteor flashes near a third-magnitude star and appears just as bright, mark it down as magnitude 3.

Estimate the altitude above the horizon (from Chapter 6) where the meteor was seen. This will help you to determine the amount, if any, of atmospheric extinction (discussed in Table 7-1). For example, if a third-magnitude meteor was seen about one third of the way between the horizon and the zenith, its altitude would be recorded as 30 degrees, with a magnitude loss of about 0.2.

Meteors can vary greatly in brightness. Some meteors may be brighter than the planet Venus (magnitude -4) and are called *fireballs*. If a fireball explodes, it is called a *bolide*. Fireballs, bolides, and regular meteors can all leave trails of smoke in the sky called *trains*. Note if the meteor leaves a train, and approximately how long the train stayed visible.

Your record sheet should have room to report all this information. You should also note the beginning and ending time of the observing sessions, as well as the date. Ruled paper allows you to run a column of numbers down the left side of the page; while across the page, you list the time of the event (in UT of course), whether the meteor was a member of a shower or a sporadic, its magnitude, altitude, and any remarks such as color or if it left a train.

You can see more meteors after midnight as you are facing the direction in which the Earth is traveling through space. Meteor observing can be a very late project, so dress warmly to view the winter, spring, and autumn (and sometimes summer) showers. The shower radiants may rise early in the evening but take a few hours to reach their highest point in the sky. The higher the radiant is, the better the meteor count and visibility—because the light from the meteors must travel through less atmosphere.

TABLE 10-1.   MAJOR METEOR SHOWERS

SHOWER	DATES	MAX.	RATE
Quadrantids	Jan 1–5	Jan 4	40
Lyrids	Apr 19–24	Apr 21	15
Eta Aquarids	May 1–8	May 5	20
Delta Aquarids	Jul 15–Aug 15	Jul 28	30
Perseids	Jul 25–Aug 18	Aug 12	50
Orionids	Oct 16–27	Oct 21	30
Leonids	Nov 13–19	Nov 17	75+
Geminids	Dec 7–15	Dec 14	60
Ursids	Dec 17–24	Dec 22	20

Meteor observing may not have the glamour of comet discovery, but it is still basic science that needs to be done. Besides, doing science from a lounge chair is plain, simple fun.

11

OBSERVING
DOUBLE STARS

*Many parts of the sky . . . are full of the most interesting and beautiful
groups and combinations.*

REV. T. W. WEBB

Astronomers have been fascinated by double stars ever since Giovanni Riccioli discovered the first one with a telescope in 1650. In the 1700s, astronomer Sir William Herschel began his investigations into stellar parallax by examining, measuring, and cataloging some 269 pairs of double stars. During the course of his work, he determined that some stars had a common orbital motion, thus discovering binary star systems. His pioneering work on the subject is considered the prime impetus for the study of double stars. Following in his father's footsteps, Sir John Herschel continued the study of double stars, reexamining his father's original list and adding 380 doubles of his own.

In 1814, astronomer F. G. Wilhelm Struve began work at the Dorpat Observatory in Estonia. Poorly equipped but with ambitious plans, Struve began a study of all the known double stars. When his initial study was completed, he had compiled a list of 795 doubles, which he published as a catalog in 1822. Having gained favor and patrons with his impressive work, Struve was able to update and modernize his equipment. During a two-year period from 1825 to 1827, Struve put his new equipment to good use, surveying 120,000 stars and discovering 3,112 new pairs. Continuing his measurements and study, he accounted for another 2,343 discoveries over the next decade.

In 1839, he became director of the newly established observatory at Poulkova. There he began his new project—the examination of all the stars in the Northern Hemisphere down to seventh magnitude. Struve was forced to turn the project over to his son Otto when the project was only a month old. Like the Herschels, the younger Struve carried on the family tradition, adding another 514 new doubles by the time the project was completed in 1942.

Professional astronomers were not the only observers to study these beautiful objects. In England a retired admiral set up a private observatory and began his own study of doubles. William Smyth, working with good optics but poor measuring equipment, compiled a list of 680 pairs from 1830 to 1843. Also in England, the Reverend W. R. Dawes began work on doubles that would span 37 years. Called "eagle-eyed" by the Astronomer Royal Sir George Airy, Rev. Dawes published a catalog in 1867 that was received as professional caliber work.

It was no different on the continent. In 1852 Baron Ercole Dembowski began to make measurements with a 5-inch telescope outside of Naples. Over the next seven years, Dembowski measured over 2,000 double stars. Between 1862 and 1878, he made over 21,000 additional measurements and received the Gold Medal from the Royal Astronomical Society for his contributions to the study of double stars.

In 1870 a gentleman from the United States entered the world of astronomy and left an indelible mark. As Clerk of the U.S. Circuit Court in Chicago, S. W. Burnham found he had ample time on his hands. The purchase of a 3-inch retractor gave him a way to fill that time. Fascinated by his newfound avocation, Burnham soon found himself drawn into the demanding study of double stars. Purchasing a 6-inch Alvan Clark retractor in 1869, he began his study of the heavens in earnest and published his first list of 81 new doubles in 1870.

His work and numerous discoveries opened doors for him throughout the world of astronomy. Burnham soon found himself welcome at observatories around the United States. During the course of his 42 years of observing, Burnham discovered over 1,200 doubles. It is a compliment to his keen-sightedness that many of the pairs he discovered and measured with his 6-inch Alvan Clark are now considered test objects for much larger instruments. In 1906, Burnham compiled his recently completed observations using the 36-inch retractor at Lick Observatory into the two-volume publication, *General Catalog of Double Stars*. This massive catalog, containing over 13,000 measurements, set the standard for double-star studies until the publication of Robert Aitken's catalog in 1932, which contained over 17,000 pairs.

Robert Aitken entered double-star astronomy in 1899, with a systematic study of stars listed in the *Bonner Durchmusterung* (BD) catalog of stars. Compiled in 1855 at the Bonn Observatory, the BD listed 458,000 stars. Aitken's project was a truly massive undertaking. For the next six years, Aitken and Professor W. J. Hussey worked on the project. In 1905, Hussey left the project and Aitken was forced to continue alone. By the time the project was completed and the final catalog published in 1932, Aitken had added 3,105 more pairs to the known list of double stars. In his six-year tenure with the project, Hussey discovered 1,329 doubles. This massive catalog is known as the *Aitken Double Star Catalog*, or simply the ADS. Another authoritative listing of double stars was published in 1984: The *Washington Double Star Catalog* (WDS), containing computerized information on 73,000 pairs.

It seems as if, to be considered an astronomer during the nineteenth century, you had to study double stars. Not really—but these studies helped broaden our scientific knowledge by leaps and bounds. From this early work came the subsequent discovery of other objects outside our solar system. It gave us an understanding of the dynamics of stars and stellar systems. It was demanding and difficult work contributed by great men of science, both amateur and professional.

THE LANGUAGE OF DOUBLE STARS

Of the 400 billion stars that make up the Milky Way galaxy we call home, about 25 percent are double stars. Called *physical pairs*, they are stars bound together by gravity orbiting around a common center. Physical pairs may be a *visual binary* system in which the component stars are visible to us. Stars like Mizar, Alpha Centauri, and Sirius are visual binaries. They may have a relatively dim companion, like Sirius and Sirius-A, but they are capable of being resolved by optical telescopes.

Another class of binary is the *spectroscopic binary*, which is composed of two stars so close together that the only way to detect the companion is by an analysis of the light reaching Earth. A final type of binary, the *astrometric binary* system, is composed of a dark body circling the primary star. This type is detected by the irregular motion of the star as it moves across the heavens, caused by the gravitational tug from the dark body. (If another civilization were looking at our Sun for about 50 of our years, they could tell that there were planets circling it—because our star scribes an irregular path against the stars.) In another type of double star, an *optical pair*, the stars simply happen to be along our line of sight from Earth and have no relationship to each other.

Of these groups, the only ones that concern amateurs like us are visual binaries and optical pairs. Studying these stars gives us a chance to follow in the footsteps of some of astronomy's great observers and to compare our visual acuity with theirs. But regardless of the depth of your interest in double stars, there are a number of terms you need to know and understand before starting any observational work.

The work of the observers just examined is reflected in the stars. The objects recorded in their catalogs retain designations noting the stars' places in the particular catalog and their discoverer. For example, Gamma Andromedae is entered as the 205th object of F. G. W. Struve's catalog. As such, it is preceded by the capital Greek letter sigma (Σ). The star is therefore also known as Σ 205.

Other designations found on star charts and in catalogs used by observers are

OΣ	Otto Struve
$\Sigma\Sigma$	Otto Struve, later catalog
H	William Herschel
h	John Herschel
Hh	J. Herschel's catalog of William's discoveries
A	Robert Aitken
ADS	*Aitken's New General Catalog of Double Stars*
β	S. W. Burnham
I	Robert T. Innes
Ho	G. W. Hough
Hu	W. J. Hussey

The stars composing the double are usually labeled A and B, with the brighter being A. In the case of a multiple star, the stars are labeled A, B, and C. During the golden age of the double star, any dim stars seen in the field of a double star were assumed to be components and, as such, were referred to as *comes* (Fig. 11-1). In many

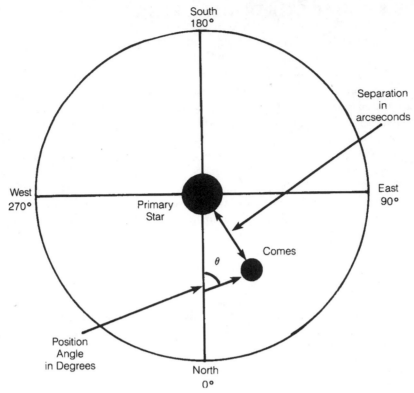

FIGURE 11-1 *Double star field through a telescope.*

cases, however, these comes were optical illusions lying along our line of sight toward the double. An example of this would be the star Aldebaran. It is a true double star with a dim, magnitude-13 companion, 31.4 arcseconds away. In the same field, however, is a magnitude-11 star at 121 arcseconds at PA 34 degrees. This star is unrelated to Aldebaran and only shares a common line of sight when seen from Earth.

The brighter star is also used as the primary in the measurement of the *position angle* (PA) of the pair. Expressed in degrees, the PA is measured counterclockwise starting from the north. If the component is west of the primary, it is said to be *preceding*; if it is east of the primary, it is said to be *following*.

The separation of the pair is measured in terms of arcseconds. Optical doubles usually have a separation of 30 arcseconds or more. In many cases, however, two stars can exceed 30 arcseconds and still be a common pair. Beta Cygni, separated by 34 arcseconds, is believed to be a physical pair, although no movement has been noticed since its discovery in 1832 by F. G. W. Struve. If two observed stars are separated by over 40 arcseconds, they are almost certainly unrelated.

OBSERVING DOUBLE STARS

The observation of a double star is dependent on a number of factors. The ability of the observer's instrument to separate the two points of light making up the double star is called the *resolving power* of the telescope. It is impossible for any telescope to make the disk of a distant star visible. It does not matter if you are using a small telescope in your backyard or the Hubble Space Telescope. No optical instrument that can show the real disk of a star, because it is too small. What you see if you throw your eyepiece slightly out of focus is the airy disk. If two airy disks are close together, they can merge into a single image. The image is considered resolved only when you can recognize those two separate disks. (Resolution limits for common amateur telescopes were listed in Chapter 2.)

The resolution limit of your telescope is set by the aperture only. Magnification has nothing to do with it. If you are observing with a 60-mm, you will never be able to split a double that is 1.5 arcseconds apart—no matter how good the seeing or how high the magnification. So if you are the proud owner of a 60-mm retractor and you head outside tonight, will you be able to see double stars down to your instrument's dawes limit of 1.9 arcseconds? No, you won't even be close.

Remember that these limits are only theoretical—in practice they are rarely reached. One reason is that both the rayleigh and dawes limits are based on a pair of stars of equal or nearly equal brightness. When you are trying to resolve an unequal pair, with one star of magnitude 3 and the companion at magnitude 9, there is no real problem if it is a wide pair. If the stars are close, however, the dimmer companion can become lost in the glare of the brighter star.

The best example of an unequal pair is the companion of Sirius. Listed as a magnitude-7 star at 10 arcseconds distance from the primary, Sirius-B would be visible if it circled any star except Sirius-A. According to our chart of instruments and their limits, Sirius-B should be easily visible in a 60-mm retractor. It isn't, because the brightness of Sirius-A literally drowns out its fainter companion. Theorized by the irregularities in the motion of Sirius, Sirius-B was discovered in 1862 by Alvan Clark as he tested his new 18.5-inch retractor. (This telescope was later purchased by the newly formed Chicago Astronomical Society and subsequently donated to the Dearborn Observatory at Northwestern University, where it is still in operation.)

Another problem comes when observing very faint pairs. It might be impossible for the eye to get enough contrast between the airy disks and the space in between. Each instrument and each observer is different. Few of us possess the sight of a Burnham or the optical excellence of an Alvan Clark telescope. The observer and his instrument form a unique personal equation that must be taken into consideration when trying to determine an observing session. The best solution would be for each individual to compile a scale that would determine exactly what pairs he or she could or could not split.

One way to measure this personal equation is for the observer to construct a *peterson diagram*. Named for Colorado amateur astronomer Harold Peterson, who developed it in the early 1950s, the diagram is an excellent way to determine how your telescope will perform on double stars. Working under the transparent skies of Durango, Colorado, Peterson determined that the main factor in resolving faint or unequal pairs is not the magnitude difference between the two stars—but the difference between the fainter component and the limiting magnitude of the instrument.

With this theory, Peterson used a 3-inch (75-mm) refractor with a 25-mm eyepiece to study 125 double stars. As he examined each double, he noted whether it resolved or not. Later he recorded his observations on a chart, plotting magnitude

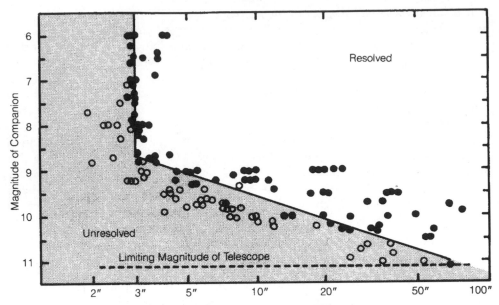

FIGURE 11-2 *Peterson diagram. (Courtesy Sky Publishing Corp., ©1980)*

along the x-axis and separation along the y-axis. Resolved pairs were recorded as filled-in circles and unresolved pairs as open circles. His completed chart showed that his telescope could resolve pairs with a magnitude-9 companion as close as 3 arc-seconds (Fig. 11-2). After that magnitude limit was reached, the detection of fainter companions required wider separation.

To get an idea of the resolution limit of your own instrument, it is a good idea to create your own peterson diagram. First, determine the limiting magnitude of your telescope. Don't go by the published numbers—take the scope out and determine the faintest star you can see. Use an AAVSO chart like the one in Chapter 7 as a guide. For uniformity, use the same eyepiece for all observations.

Observations should be made when seeing is good. Peterson used the visibility of the companion of Polaris as a gauge of seeing. Doubles should be well placed, prefer-ably on or near the meridian. Using a list of doubles, record whether or not you can split the stars. When you have accumulated a number of observations—about 100 should suffice—transfer your results to a graph. With your own diagram complete, you will know instantly if a double star is within reach of your telescope.

Once you have determined the peterson limit of your instrument, it is only human nature to try to exceed it. Of the three limits we have discussed here—the rayleigh, the dawes, and the peterson—the first two are based on the unchangeable fact of instrument aperture and cannot be exceeded. The peterson limit is a standard drawn against one set of circumstances. It assumes that the observations you made to construct the chart were made at your normal observation site. It also assumes that all observations were made with the same eyepiece—one with a low magnification. Because the peterson limit is a function of the telescope's limiting magnitude and the magnitude of the fainter member of the double-star system, you can try to exceed it—and you may succeed.

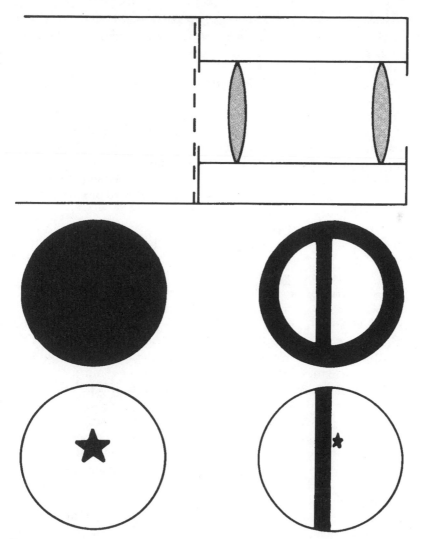

FIGURE 11-3 *Occulting bar.*

There are a number of strategies you can use when trying to split a very close double. One of the easiest is the use of an occulting bar at the field stop of the eyepiece (Fig. 11-3). To construct one, you need a strip of thin, dark material that is easily cut or shaped. A sheet of exposed film is good; just cut a circle in it the size of the inside of your eyepiece barrel. Then shape it until you have a small strip, 25 mm long and about 5 mm wide. Finally, cement this to the field stop of the eyepiece. To use the bar, rotate the eyepiece until the brighter component is hidden behind the strip. The fainter close companion should become visible when the glare of the primary star is cut by the film.

If you are using a retractor, another way to get a look at the close, dim companion is to make use of the diffraction pattern reflector owners take for granted. The pattern is created when light passes around the four vanes that hold a reflector's secondary mirror in place. It shows up as four *spikes* that are seen around the star. The light from the star is channeled into the spikes, leaving four spaces in which a dim object can be seen. Because the retractor has an unobstructed light path, you have to place something in the way of the light to get the diffraction pattern such as a diffraction mask. The easiest way is to criss-cross two pieces of duct or electrical tape across the end of the dewcap (the piece that moves up and down over an eyepiece). *Do not place the tape on the lens.*

Now the light entering the tube will be diffracted by the tape, and you will see four spikes around the brighter component. The only drawback to this method is that the tape cannot be moved around easily, so the double components must be placed so that the fainter star is in the space created by the diffraction pattern. You can remedy this by using some ingenuity; place a piece of cardboard or plastic sleeve over the end of the dewcap to which you attach the tape.

Some observers recommend a tape width of 10 mm for acceptable results, but each observer must determine his or her own tape size through experimentation. Make up a variety of strips and try them on various types of doubles. Is there a difference when using a wide band on a very close double? A narrow band? Keep track—nothing is ever a waste of time in science and astronomy.

A similar method makes use of a special type of aperture mask (Fig. 11-4). Using tape, film, or even cardboard, create a hexagonal (six-sided) mask to fit over the end of the dewcap or telescope tube. Use a sleeve so that you can rotate the mask. This hexagonal mask will give you a diffraction pattern of six spikes around a bright star. The sleeve allows the pattern to be rotated—so that the faint close companion is easier located in the glare of the primary. Like the four-spike cross, it is best to experiment with a variety of sizes for the mask.

The average eye can resolve a pair of stars 4 arcminutes apart. A star like Epsilon Lyrae, at slightly over 3 arcminutes in separation, appears as an elongated star to the naked eye of most observers, although the sharp-eyed can distinguish two stars. Looking at Epsilon with a pair of 7 × 35 binoculars clears up any doubt about its duplicity: The two stars stand clearly separate.

The components that make up the Epsilon Lyrae "double double" are much more difficult to separate. Epsilon-1 is the wider of the two systems, with a separation of 2.8 arcseconds. Epsilon-2 is separated by 2.2 arcseconds. Both stars are well within the limits of both a 60-mm and an 80-mm retractor. Also, they transit the

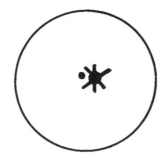

FIGURE 11-4 *Aperture mask.*

meridian high in the sky, and the stars are about equal in brightness. All of these things combine to create a unique test object for the small telescope. On a night with good seeing, begin with a low-power eyepiece and work your way up in magnification. Which eyepiece allows you to see the elongated images of both stars? Which eyepiece actually resolves Epsilon-1? Which resolves Epsilon-2? For example, with an 80-mm and good seeing, you can resolve Epsilon-1 with a 12-mm ortho. Epsilon-2 takes a bit more, with a 9-mm sufficient for the entire four-star system.

When trying to resolve a close pair, you must remember that the eye can resolve two objects 240 arcseconds apart with no assistance. Using 240 arcseconds as a threshold of visibility, divide that figure by the separation distance of the pair to determine an approximate magnification for resolution. At 2.7 arcseconds, Epsilon-1 Lyrae should need a magnification of about 88 for resolution; Epsilon-2, about 110. An object like Gamma Arietis with a separation of 7.8 arcseconds needs a magnification of about 30. Use this formula as a guide—but remember it is only a guide.

Magnification is dependent on both stable seeing and a solid mount. At high magnification, the field of the telescope seems to speed past your eye so fast that it is disorienting. Remember, as the telescope magnifies an image, it also seems to increase the speed at which the Earth rotates. You must have a very solid mounting to be able to concentrate on a pair at high power. The best mounting is an equatorial equipped with a drive, but a smooth altazimuth mount can be used with a steady hand.

DOUBLE STAR COLORS

Observers in the eighteenth and nineteenth centuries were fascinated by double stars and the colors of the components. The colors described by Admiral Smyth, Rev. Webb, and other observers were so beautiful, they brought tears to the eyes of some of their readers. Imagine the contrasting colors of apple green and cherry red given by Smyth for 95 Herculis. The elder Struve went as far as developing his own color scale, including what he called pinkish-green, when recording his observations. These astronomers thought that by observing the colors of the stars, they could gain insight into their makeup. For many years these reports were scoffed at and written off as overactive imaginations on the part of the observers. Now we are finding they were partially correct.

In many cases, the colors we see when we observe a pair are caused by the reaction of the eye's rods and cones to the two stars. But there does seem to be a correlation between reported colors and spectral type. Research into this subject was conducted by four amateurs using the 24-inch reflector at the Mt. Cuba Observatory in Delaware. After studying some 200 pairs, they found a strong correlation between color, spectral type, and surface temperature. They further found that pairs with inconsistent color reports were those with a wide separation. The closer the pair, the more consistent the color reports.

It is therefore essential that you record the apparent colors of the stars you are observing. Not only are the colors pleasing to the eye, but they can tell you quite a bit about the makeup of the pair under observation. Stars that appear red, for example, are usually members of spectral class M, while stars seen as blue or with blue tints belong to spectral classes O through A. Remember, these conclusions were reached by amateur astronomers like you. They had access to professional equipment, but they gained that access by taking their astronomy seriously. So don't let anyone tell you that the days of serious contributions by amateurs are over.

MAKING MEASUREMENTS OF DOUBLE STARS

If you are interested in observing double stars, what can you do other than look at them? For a start, you can measure their separations and estimate their PA. Don't hesitate—even if you don't have a fancy micrometer or the money to buy one. Don't worry. Most of us don't have access to precision micrometers either, but there are other ways to measure doubles.

The easiest way to make measurements is with a micrometer reticle eyepiece. Available from a wide variety of companies, one can cost between $70 and $150. In addition to its use with double stars, it can help measure planetary disk size and determine CM for transit timings, so it's a handy item to have around. The reticle eyepiece is built around a piece of optical glass. A cross-line pattern is scribed onto the glass and is broken into a number of major divisions, which are further divided into a number of minor subdivisions. The glass fits onto an eyepiece, usually from 12 mm to 6 mm in focal length, at its focal plane. Most of these reticles include a red LED that illuminates the reticle scale and makes it easy to see against the dark background of the sky.

Before you use the reticle eyepiece, you have to calibrate its subdivisions. It is helpful to know just how many arcseconds each one covers. Mark Stauffer, an amateur living in Baton Rouge, Louisiana, used a reticle eyepiece and recommends the use of a wide double for this calibration. Using a Meade reticle eyepiece with a focal length of 12 mm on his Celestron C90 telescope, Mark determined that each of the divisions scribed into the reticle is equal to 8 arcseconds. This figure is going to change for each telescope, so you must calibrate it for your own system. Mark used Epsilon Lyrae, which has a wide separation of 208 arcseconds.

To do the calibrations, adjust the reticle so that the micrometer line passes through both major components, then simply count the number of divisions between them. For Mark's C90, this worked out to 26 divisions. If you divide the measured distance in arcseconds (208) by the number of divisions (26)—you get 208/26 or 8 arcseconds per division. This will be different for each telescope, so check yours.

Mark further tested his reticle by measuring the separation of the double star Gamma Arietis with the two stars covering one division of the reticle. This gave an observed separation of 8 arcseconds—with the actual separation at 7.8 arcseconds. However, a micrometer reticle eyepiece is not able to tell you the PA of the pair. This will have to be estimated by the eye.

Another, more accurate device for determining the measurement of a double star is a diffraction grating micrometer (DGM). The principle behind the DGM is that when the image of a star passes through a series of closely aligned slits, it is seen as a bright point source surrounded by a number of satellite images (Fig. 11-5). The

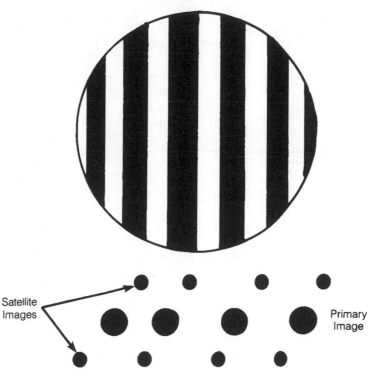

FIGURE 11-5 *The DGM grating and the type of image it produces.*

total width of the bar and slit (p) can be changed so you can vary the distance between the primary and satellite images (z). By varying z, which will be different for each observer, a variety of gratings can be made, each tailored for a specific separation distance. The value of z used for a pair should be close to, but not larger than, their actual separation. Because each person has a different z value, see (Fig. 11-6) for a method you can use to determine yours. The actual measurements of the double star are made by rotating the grating so that the images of the stars form a right triangle. A reading of the angles involved is made from the micrometer. The separation and PA can be computed from the average of a number of measurements. A DGM can be constructed at very little expense and can be used with any type of telescope. Mark Stauffer built and used a DGM with his 80-mm f/11.4 retractor and found the results a great improvement over the micrometer reticle eyepiece.

To make a DGM you will need some poster board, a couple of protractors, and a sharp knife. Cut a circle from black poster board that corresponds to the diameter of the telescope's dew cap. Then draw a series of lines that correspond to the value of p you selected. Next fit two protractors onto the edge of the completed grating, aligning the zero point of the protractors to the horizontal position of the grating. The appearance of the unit can be improved by making the protractor/micrometer scale fit flush along the barrel of the dew cap. This can be done easily by using tape, a ruler, and graph paper. Mark mounted his DGM by making a ring from a 1-inch-

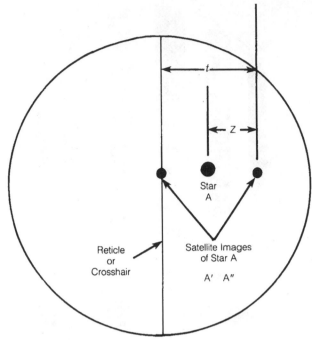

FIGURE 11-6 *To determine your individual z factor, find a double with a high declination and time the images as they move across a reticle or crosshair. When you have the time, use z = (7.5 × time) × the cosine of the star's declination.*

wide-by ⅛-inch-thick aluminum strip. He used the lip from a margarine tub lid to attach the entire assembly to the telescope.

The accuracy of the DGM depends on the separation of the pairs being measured. The ability of the telescope to resolve close pairs is crucial to the accuracy of the DGM because it can give a built-in error factor for the measurements attained with any size telescope. When he used the DGM with his 80-mm telescope on close pairs, Mark determined that the DGM is subject to theoretical errors of up to 30 percent. As the separation of the pairs increases, there is a corresponding decrease in the theoretical error to about 7 percent. This means that you can more accurately measure wider pairs. In actual use Mark determined that for pairs of about 10 arcseconds separation he had an error of 3 percent, while for pairs of 5 arcseconds separation the error increased to 10 percent.

So what do you do with this DGM? An excellent project would be to examine and measure some of the wider pairs listed in a catalog like the *Sky Catalogue 2000*. Professional astronomers are interested in the properties of doubles, but they tend to limit their interest to very close visual pairs, spectroscopic binaries, and astrometric binaries. Their interest in wide visual pairs is limited to taking measurements, computing their orbits, and possibly checking an ephemeris containing predicted separations and position angles. There is very little follow-up on the accuracy of these predictions, especially for wide pairs. An amateur using a DGM could measure and test this accuracy very easily.

DOUBLE STARS FOR THE TELESCOPE

There are so many double stars in the sky that it is difficult to come up with an all-inclusive list. Over the years, Mark Stauffer developed a list of double stars that shows the wide variety of colors and separations found in these objects. All of his observations were made with an 80-mm f/11.4 retractor. Mark's list is a good starting point for your own observations because each of the stars is easy to find. (His comments follow each entry.)

Beta Cygni (Alberio) Gold and blue; best seen at low power. One of the finest pairs in the sky.

Gamma Andromedae A fine yellow and blue double. Just as nice as, if not nicer than, Alberio. Best viewed at moderate magnification.

Gamma Arietis A lovely pair of equal magnitude stars that look like a distant pair of headlights.

Epsilon Lyrae The renowned "double double." I've observed this remarkable pair of pairs at 105× and was able to split both pairs of stars, provided the seeing was good.

Zeta Ursa Majoris (Mizar) An easy pair of magnitude 2 and 4 stars that resolve at 35×. A very nice system and field too, with Alcor nearby at about 12 arcminutes. A faint star is located between Mizar and Alcor off to the side.

Alpha Ursa Minoris (Polaris) A nice unequal pair at magnitudes 2 and 9 separated by 18 arcseconds. Look very carefully for the bluish-colored comes.

Xi Bootes A wonderful color-contrasted double that is also a binary star. The primary is yellowish while the companion sometimes appears orange, sometimes reddish-purple. A fine pair to observe at moderate to high power.

Beta Orionis (Rigel) A fine unequal pair at magnitude 0.1 and 6.8. The companion is easily observed at 105×.

Gamma Delphini An easy pair of yellow stars about 10 arcseconds apart. Close by is another pair (Struve 2725), which are resolvable at 35× when the seeing is good.

Zeta Lyrae A very nice contrasting pair of yellow and blue suns. An easy pair at low power.

Omicron-1 Cygni A very wide color-contrasted triple with one companion at 107 arcseconds and the other at 338. Both are most likely optical companions. Can you tell that one of the companions is bluer than the other? Which one is it? A really nice sight at low power in a small scope.

Lambda Orionis A close pair of yellow and light blue stars worth viewing at moderate to high magnifications.

Theta Orionis (The Trapezium) An excellent quadruple found within the Orion Nebula (M42). An easy target for small telescopes at any magnification.

Epsilon Bootes A difficult pair for the small scope, but I was able to split the yellow-orange primary and greenish companion with 182×.

Iota Cassiopeiae A fine close triple of magnitude 5, 7, and 8 stars. The magnitude 5 and 7 stars are separated by 7 arcseconds, while the magnitude 5 and 8 stars are 2 arcseconds apart. Best observed at high magnifications. Some color to this system: white primary, bluish companions.

Eta Cassiopeiae A real jewel of the night sky. A binary with a yellowish primary and a companion whose color I've seen to be orange, red, even purple. A double worth observing at any magnification.

Epsilon Pegasi (Enif) A very wide pair of orange and blue stars. A fine object to view at low power.

Alpha Herculis (Rasalgeth) A wonderful double composed of orange and greenish-blue stars separated by about 5 arcseconds. Best observed at moderate and higher power.

Beta Monocerotis A very nice triple. The two companions themselves form a close duo. The entire triple can be best seen at moderately high power to take advantage of the close pair formed by the two companions.

Alpha Geminorum (Castor) A truly fine binary composed of magnitude 2 and 3 stars, just resolved at 105×.

Gamma Leonis A close pair of yellow stars that is within reach of a small telescope at moderate magnification. A pretty sight.

12

OBSERVING VARIABLE STARS

. . . approach the observing of variable stars with the utmost caution. It is easy to become an addict . . .

LESLIE PELTIER

That is probably the most accurate statement you will ever hear in amateur astronomy: Variable star observing is habit-forming. Variable star observers are among the most active and dedicated members of the amateur astronomical community.

By 1999, more than 550 observers—over half of whom live outside the United States—in over 40 countries contributed 350,000 observations of variable star magnitude estimates yearly to the American Association of Variable Star Observers (AAVSO). That averages out to 500 estimates per member. In fact, the AAVSO states that about 6,000 observers have contributed over 9 million observations of variable stars to the AAVSO International Database since the founding of the AAVSO in 1911. These people are very dedicated.

The reason so many observations are made by such a small group of observers is simple: These people know that their observations are going to be shared with professional astronomers who need this information. Even the people who were only able to contribute one observation know that the single observation is significant because it, all by itself, fills a need.

TYPES OF VARIABLE STARS

The stars that instill so much dedication in their observers change their brightness for one of two reasons. If the change is caused by mechanisms within the star itself, the star is called an *intrinsic variable*. Stars that expand and contract with some regularity, stars that "hiccup" during their life cycle, and stars that literally blow themselves apart are all examples of intrinsic variables.

If the change takes place because an external event, the star is called an *extrinsic variable*. Examples are two stars revolving around a common point, with one blocking the light headed toward Earth, and stars covered with massive *starspots*.

Variables are identified by a letter designation. The first variable discovered in a constellation is always labeled *R*, followed by *S*, then *T*, and so on, beginning over with *RR* when the last letter is reached. Variables also carry a six-number designation known as the *Harvard designation*, which is simply the coordinates of the star for the year 1900. The first four numbers denote its right ascension in hours and minutes, and the last two denote its declination. If the star is south of the celestial equator and will have a negative declination, the last two numbers are underlined or preceded by a minus sign.

For example, the first variable discovered in the constellation Cancer carries the name *R Cancri* and the Harvard designation *081112*. The first variable discovered in Aquarius is *R Aquarii* and has the Harvard designation of *2338-15*, denoting that it is south of the celestial equator.

Each of these primary classes is broken into a number of variable subclasses. Each subclass includes stars that share similar brightness variation profiles, called *light curves*. The curves are plots of brightness against the time it takes for the brightness to complete a cycle, or *period*. Most of these subclasses are named for the first or most famous star in the group.

Cepheid variables are named after Delta Cephei. They show a small brightness range of 1 to 2 magnitudes, and periods of from 1 to 70 days. Their light curve (Fig. 12-1) shows smooth growth and decline in brightness. Because their interior changes cause them to expand and contract, these stars are members of the intrinsic variable group.

RR Lyrae stars also pulsate, but they are much older stars than the cepheids. Named after the star RR Lyrae, their periods are much faster—never more than a

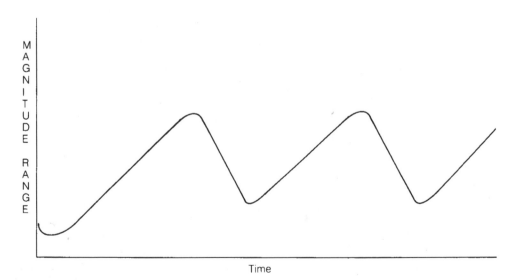

FIGURE 12-1 *Light curve of Cepheid variable.*

day—and their brightness range is never more than two magnitudes. Their light curves (Fig. 12-2) show sharper peaks and steeper valleys than the smoother cepheids. Called *cluster variables*, many RR Lyrae stars are found in globular star clusters. They are members of the intrinsic group.

RV Tauri stars, named after the first star to be discovered with these characteristics, show as much as a three-magnitude range in brightness. Their periods, measured from one point of minimum brightness*(minimum)* to the next, can be from 50 to 150 days long. Their light curves show irregular brightness peaks and another minimum, not quite as dim, between the deeper minima. These stars, which behave so strangely, are also members of the intrinsic group.

Long-period variables (LPVs) are the largest subclass within the intrinsic group. They have well-defined periods that last from 75 to 1,000 days and a brightness range that can exceed five magnitudes. Their light curves show smooth rises to levels of maximum brightness *(maxima)* followed by equally smooth drops to minima.

Related to the LPVs are the *semiregular variables* (SRs). These stars can't make up their minds. In some years they show regular maxima and minima; in others they show no variation in brightness. Another group of these indecisive stars are the *irregular variables*, which show little evidence of having any kind of period.

All of these variables have something in common: Their variable brightness is caused by expansion and contraction, a pulsation of the stars themselves. These types of stars are called *pulsating variables*.

The next group of intrinsic variables, called *eruptive variables*, have even more dramatic internal changes. They are all subject to massive eruptions of brightness from blowing themselves to pieces or just blowing off their outer layers.

The extreme case of the eruptive variable is the *supernova*. A supernova can brighten by as much as 20 magnitudes—an increase of 90 million times its original

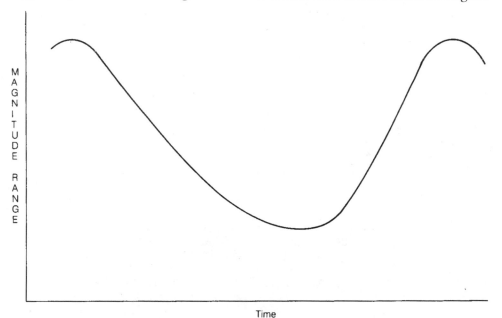

FIGURE 12-2 *Light curve of RR Lyrae variable.*

brightness—in a very short period. Then it gradually fades to a mere fraction of its original brightness, until it is no longer detected.

The smaller version of the supernova is the much more common *nova*. Novae can brighten by seven or more magnitudes in a matter of days before they gradually fade to their original magnitude. Novae usually only brighten once during their lifetime, but another class, the *recurrent novae*, can undergo two or more outbursts during their life.

U Geminorum (U Gem) stars are also related to novae and recurrent novae. These stars can increase in brightness by two to six magnitudes, but they do so on a fairly regular basis at intervals of a few months. They return to their normal brightness much more rapidly, usually in about a week. Related to the U Gem stars are variables in the Z Camelopardalis (Z Cam) family. Like the U Gem stars, these are subject to rapid outbursts, but they can get stuck at a certain brightness as their brightness declines. Z Cam stars are also called *dwarf novae*.

UV Ceti stars, also called *flare stars*, also undergo extremely sudden and rapid increases in brightness. They can increase by up to six magnitudes in a matter of minutes.

Finally, we have the last of the intrinsics, the R Coronae Borealis (R CrB) stars. Unlike the other eruptives, R CrB stars maintain a certain level and then decline in brightness. They can decrease by as much as nine magnitudes before returning to their former magnitude over a period of a few months to several years—in a way, a reverse nova. *Eclipsing binaries* are members of the extrinsic group. The cause of their variability is simple: The plane of the binary star system lies along our line of

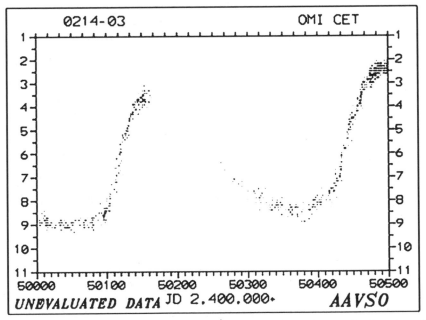

FIGURE 12-3 *Light curve chart for Mira (OMI CET), the first variable star discovered and prototype for Mira type variables (do not use this chart for official research purposes—it has not passed AAVSO's strict quality control). (Courtesy AAVSO)*

sight. As star A and star B swing through their orbit, one blocks the light of the other and we see a drop in brightness at a regular interval. Beta Persei, also known as *Algol*, is an excellent example of such a system.

Although the amateur has a wide variety of variable types to choose from (Figs. 12-3, 12-4, and 12-5), it is easier for the new observer to get his or her feet wet by concentrating on long-period and semiregular variables. So those are the types we are going to talk about here.

FINDING VARIABLE STARS

Finding variable stars is the hardest part of observing them. Most amateurs are used to looking for faint objects in their telescopes, but variables are different. When you are looking for a double star, your subconscious mind knows you are going to see two stars close together. The description of the double with its colors, separation, and position angle help the eye and the brain find the stars. When you are searching for a faint deep-sky object, your subconscious mind has the same advantage. It will give you certain visual clues about your target. It knows that whatever you are looking for has a certain size, shape, and configuration. It allows you to scan the star field and disregard the dozens of star images that don't fit the pattern.

So how do you visualize a variable star? It's not fuzzy. It's not a mass of stars, but a single star. It's not normally found with another star close by at a particular posi-

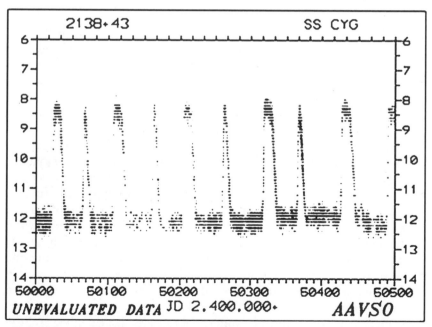

FIGURE 12-4 *Light curve chart for SS CYG, the prototype of the cataclysmic variables, is also the most popular star in the AAVSO observing program (do not use this chart for official research purposes—it has not passed AAVSO's strict quality control). (Courtesy AAVSO)*

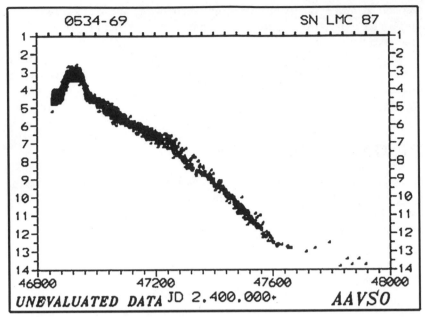

FIGURE 12-5 *Light curve chart for Supernova 1987a in the Southern Hemisphere's Large Magellanic Cloud (do not use this chart for official research purposes—it has not passed AAVSO's strict quality control). (Courtesy AAVSO)*

tion angle. When you are looking for a variable star, there are no real visual clues. You are trying to pick one star out of a background of many. How do you do this?

The first thing to do is learn about the general area through which you will be searching. AAVSO helps with this task by publishing a series of charts for each variable star (Fig. 12-6). On each chart is the basic information for the variable: its name and Harvard designation, its magnitude range and period, and sometimes information about its spectral type or color. AAVSO charts come in seven versions, each keyed to a specific field-of-view size and each having a successively smaller area of coverage. The chart type is listed next to the box containing the star's Harvard designation. Each chart contains the variable, which is usually indicated by an open circle, along with a series of stars and their magnitudes so that you can make brightness comparisons.

Type (a) charts cover an area of 15 square degrees on a scale of 1 millimeter (mm) equal to 5 arcminutes. The limiting magnitude or faintest star shown is magnitude 7.5. These stars are listed and printed with north at the top of the chart. With this orientation, type (a) charts are excellent for use with binoculars. You will notice that the charts list magnitudes as whole numbers. A magnitude 6.5 star is listed as 65 with the decimal point eliminated. This is done so that you won't confuse the decimal point with a star.

Type (ab) charts have a scale of 1 mm equal to 8 arcseconds. They are issued for brighter stars and have a fainter limiting magnitude. Because they are also oriented with north at the top, they are useful for variables that can be seen with binoculars.

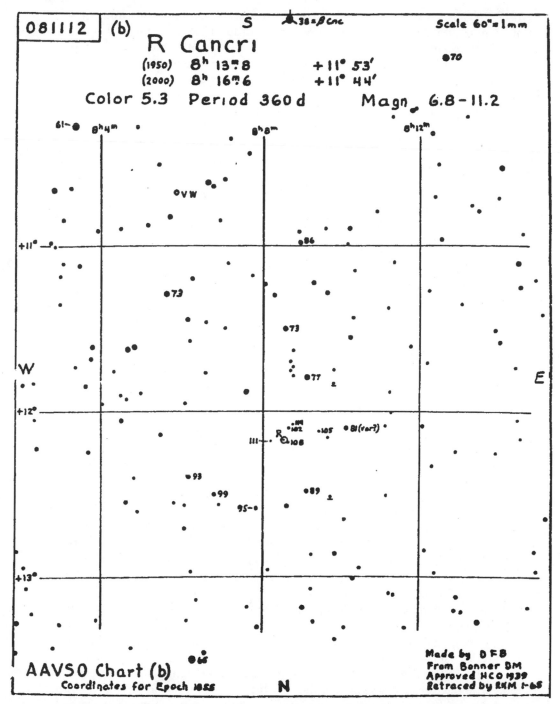

FIGURE 12-6 *AAVSO variable star chart for R Cancri (R Cnc). (Reprint by special permission of the American Association of Variable Star Observers, through its director, Janet Mattei)*

Type (b) charts cover an area of 3 square degrees and have a scale of 1 mm equal to 1 arcminute. They contain stars as faint as magnitude 11 and are the first in the series to be printed with the correct astronomical orientation of south at the top. Type (b) charts are recommended for use with telescopes up to 3 inches in aperture.

Type (c) charts have a scale of 1 mm equal to 40 arcseconds and are printed with south at the top. They are intended for use with telescopes from 3 inches to 6 inches in aperture.

Type (d) charts have a scale of 1 mm equal to 20 arcseconds and are printed with south at the top. They are for telescopes 6 inches and larger.

Type (e) charts have a scale of 1 mm equal to 10 arcseconds and are for use with faint variables. Designed for use with telescopes 6 inches and larger, they are printed with south at the top.

Type (f) charts have a scale of 1 mm equal to 5 arcseconds. They are used when the variable is faint and the field contains many other faint stars.

AAVSO also publishes a number of *Finder Charts* for various areas of the sky. These cover a 10-square-degree area at a scale of 1 mm equal to 6 arcminutes. They are designed to help locate the field of the variable and contain no magnitudes, so they are not to be used for making brightness estimates. A number of charts are also available for those with telescopes that have reversed fields, such as retractors and catadioptrics with star diagonals.

You can use an (a) or (ab) chart to locate the target variable at the telescope and then use it or another chart to estimate the star's brightness. Not every variable on AAVSO's list is listed on a complete range of charts, however. Some of the brighter variables that can be seen through binoculars have only the (a) and (ab) charts, just as some of the fainter variables have only one or two charts.

Regardless of the number and type of charts available for your target variable, it is best to first locate the star in your own star atlas. If the star is too faint to be listed, use the coordinates given at the top of the chart and carefully plot the star yourself with a pencil. This lets you get a look at the big picture of the area in which the variable is located. Once you have the variable located or plotted, look around it for bright stars that will help you zero in on the variable. Use the ring you made with a coat hanger to get an idea of what the fields will look like in the finder.

Once you feel comfortable with the general area, take out the AAVSO charts for the variable and try to find visual clues that will help you find the star. In most cases there won't be any for you to use, so you are going to have to devise your own visual clues. Look on the chart for *asterisms* or groups of stars that form a recognizable shape. These will be your landmarks when you are at the eyepiece. For example, if you try to find R Cancri (Fig. 12-7), look for three landmarks, especially when the star is faint. First look for the little squiggly line of stars (A) just southeast of the variable. When you find this, you know you're close. Then look for the variable to form a kite shape with three other stars (B) in the field. The final landmark is the arc of magnitude-9 stars (C) just north of the variable. This way, you can check your position not once but three times, so you're sure you're in the right place.

Once you have reached the area of the variable, you still must be careful. If you can see the variable in relation to the other stars, everything is fine. But when a variable is faint it can easily be confused with some of the comparison stars listed on your chart. Take some time to carefully check the position of each plotted comparison

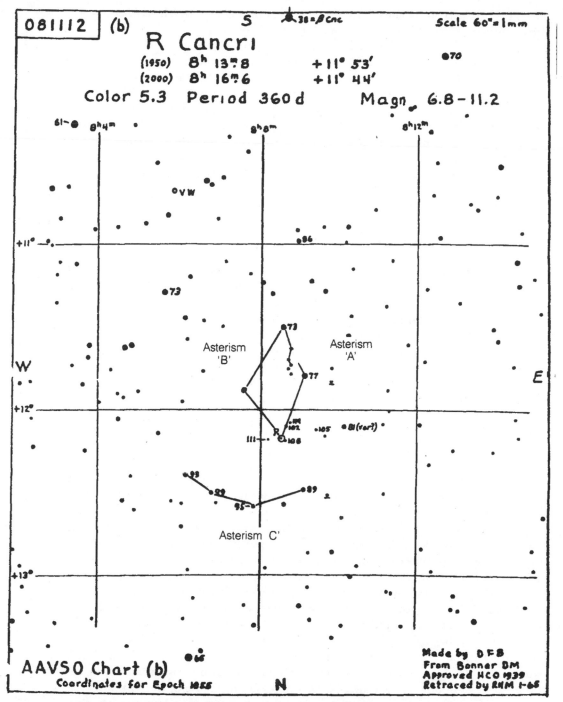

FIGURE 12-7 *R Cnc chart with asterisms marked. (Reprint by special permission of the American Association of Variable Star Observers, through its director, Janet Mattei)*

star. This might seem tedious, but it's better to be sure about which star is the variable than to make six magnitude estimates of the wrong star.

This entire process might put a damper on your enthusiasm for variable star observing before you start—but don't let it. The first time you add a variable to your observing list, these precautions will make a lot of sense. You are venturing into a new area of the sky—an area that might not be familiar to you. The next time you observe the variable, your subconscious mind will know some landmarks because you've been over this ground before. Each time you see the area, you'll find the variable a little faster. And one day, all you'll have to do is point the telescope and the variable will be there—or so it will seem.

MAKING MAGNITUDE ESTIMATES

Now you are there. You've located the proper field, your eye is glued to the eyepiece, and you find yourself looking at a field of stars. What happens next? The desired end result of variable star observation is an accurate estimate of the star's brightness on the night of the observation, or a *magnitude estimate*. To do this, simply compare the brightness of the variable with the brightness of known comparison stars that never vary their light output.

Remember the way to identify each of the comparison stars we presented earlier? This is where it pays off. First, identify the variable. Many variables, particularly the LPVs, appear red in color. That red color may be your first clue to finding the right star, but don't expect the variable to pop out as a red star. Check the AAVSO chart and compare the positions of each star listed on the chart with what you see in the field of view. If the star is faint, you might have to switch to a higher-power eyepiece to bring out the fainter stars. If you do that, get out the next chart in the variable's series. If you use the (b) chart (Fig. 12-7) to zero in on R Cancri and can't quite make out which star matches the variable, switch to the (c) chart (Fig. 12-8). It will give you a better idea of how the other stars in the field are positioned around R Cancri. And if you've prepared well, you can use the asterisms you've marked on that chart to better refine the position of R Cancri.

If you have checked everything in the field and the variable still eludes you, chances are it is at a brightness below your telescope's limiting magnitude. Check the field once more and make a mental note, followed by a written one, as to which is the faintest star in the field that you can see. Indicate this with a notation such as *R Cnc not visible—(108*. This indicates that on the night of your observation R Cancri (R Cnc) was fainter than a star of magnitude 10.8, the faintest star you could see. The open parenthesis indicates "fainter than" and is a valid, and valuable, magnitude estimate.

If you find the variable, it is time to make a direct comparison. Let's say you find your target variable and it appears fairly bright. How do you proceed? First, decide which AAVSO chart you are going to use to make your comparisons and estimates. If the variable, in our case R Cancri, is very apparent in the field, you can safely use the (b) chart. Begin by letting your eye get used to the scene for a few minutes, then pick a star you feel is close to R's brightness. Let's say it's the magnitude-93 star at the

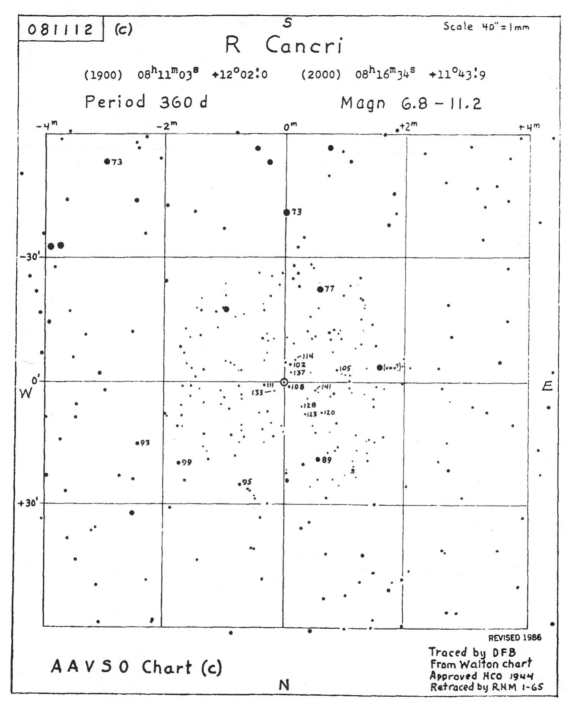

FIGURE 12-8 *Type C R Cnc chart. (Reprint by special permission of the American Association of Variable Star Observers, through its director, Janet Mattei)*

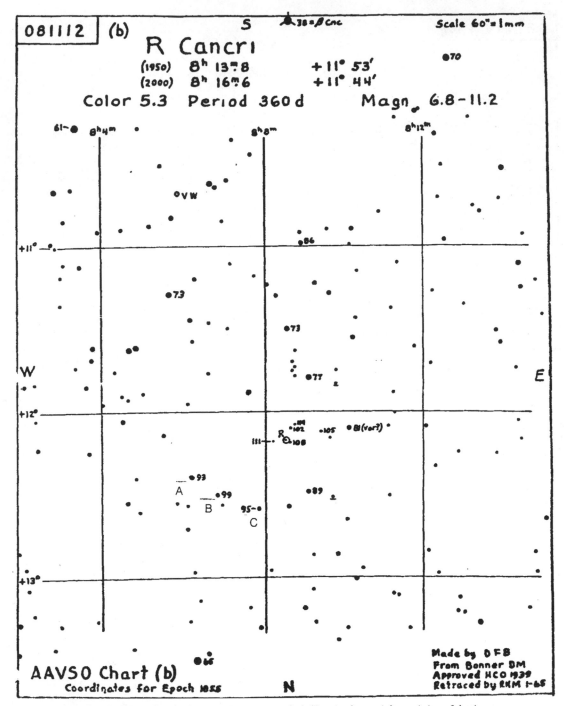

FIGURE 12-9 *Type B R Cnc chart with comparison stars marked. (Reprint by special permission of the American Association of Variable Star Observers, through its director, Janet Mattei)*

west end of the arc. Look at the two stars and quickly move your eye between them. Is R brighter or fainter than A? Fainter? Move north along the arc one star to star B at magnitude 99. Is R brighter or fainter than that one? Brighter? Now you have determined that R Cancri is somewhere between magnitudes 93 and 99 (remember the decimal point is left out).

Are there any stars of this magnitude range that you can compare with R? Yes, there is: Star C, the next star along the arc, is magnitude 95. Is R brighter or fainter than star C? After a few minutes you decide that R is slightly fainter than C, but brighter than B. Now it's time to compare all three stars—B, R, and C. Move your eye from B to R to C to R then back to B. Try to place where R's brightness falls. Does R appear to be closer to the brightness of one or the other? If it appears closer to C than to B, your estimate would be magnitude 96. If it appears to be closer to B than C, your estimate would be 98. If it appears to be midway between the two comparison stars, you would estimate R at magnitude 97. When you make your final estimate record it in your log. The entry would look something like this: *R Cnc - 96. 93/95/99.*

This means your estimate of R Cancri was magnitude 9.6 and you used three stars at magnitudes 9.3, 9.5, and 9.9 for the comparison. It is important to list the comparison stars you used, so be sure to mark them down (Fig. 12-9).

The trick is to use as many comparison stars as you can to narrow the range of brightness. By doing this, you are bringing the final estimate to within a narrow range of possibilities and improving its accuracy. If you find that there are only two comparison stars, use the two stars. But if you can, try to find a third, especially if you are making an estimate on a star for the first time.

You need to be careful when estimating the magnitudes of variables that appear to be red in color because of the *purkinje effect*. Red stars will excite the retina because the retina does not react to them as fast. Because of this, the image builds up on the retina and sends a stronger signal to the brain than other, nonred stars. Consequently, red variables are seen as being slightly brighter. This can produce an incorrect magnitude estimate. When you are looking at a red long-period variable, try not to concentrate too hard on the star for any length of time. Keep your eyes moving around the field to minimize the amount of time the variable stays on any single spot on the retina. Another way to overcome the effect is to throw the field slightly out of focus to reduce the intensity of the red color.

KEEPING RECORDS

Once you have made all your estimates for the evening, it is time to put them in order. You want to use a system that keeps everything handy, but one that won't influence your estimates at the telescope. A good way to do this is to keep all your observations intact in your observation log, with one page for each night's work. Once inside, transfer this information to a master log, allowing one sheet for each variable star.

Looseleaf ledger sheets with six columns work well for this master log. You can head each sheet with information relevant to the variable and label the columns of the sheet with headings such as: remarks (the wide column), date, time (local), time (UT), estimate, comparison stars, and conditions. If you happen to have a personal computer, a database or spreadsheet program works nicely.

TABLE 12-1.   CONVERSION TO DECIMAL DAY FOR OBSERVATIONS MADE BETWEEN

LOCAL TIME

EST	CST	MST	PST	AND	EST	CST	MST	PST	DECIMAL
324	224	—	—		548	448	—	—	0.4
549	449	349	249		811	711	611	511	0.5
812	712	612	512		1036	936	836	736	0.6
1037	937	837	737		1259	1159	1059	959	0.7
1300	1200	1100	1000		1524	1424	1324	1224	0.8
1525	1425	1325	1225		1747	1647	1547	1447	0.9
1748	1648	1548	1448		2012	1912	1812	1712	1.0
2013	1913	1813	1713		2235	2135	2035	1935	1.1
2236	2136	2036	1936		—	—	—	2200	1.2
—	—	—	2201		—	—	—	—	1.3

(Adapted from AAVSO, Janet Mattei, Director.)

If you are submitting estimates to AAVSO, get in the habit of listing the date according to the Julian Day (JD) calendar. Julian days are listed consecutively from the starting date of January 1, 4713 B.C. July 31, 1988, for example, is JD 2,447,374. This gives AAVSO a scale against which to plot light curves and make its predictions for variables.

This table is used to convert observation times of variable star estimates from local time to decimal fractions of a Julian Day. First convert the time of your observation to a 24-hour clock starting at noon. Three A.M. CST now becomes 1500 hours local time. Look at the table and see what two limits 1500 falls between (CST 1425 and 1647), for which the decimal day of your observation becomes 0.9. Add this figure to the Julian day of your observation.

The big difference is that JDs are measured from noon to noon instead of midnight to midnight, so if you insist on using our calendar, label observations made on the night of July 30 as July 30-31. AAVSO can supply you with the current year's JD calendar. A number of commercially available astronomical calendars list the JD as well as the date you are used to seeing. AAVSO also requires that observations be listed with the hour in decimal form. Table 12-1 gives you conversions for changing UT and local time into decimal hours.

To submit your estimates to AAVSO, you need only fill out the organization's monthly form (Figs. 12-10, 12-11, and 12-12). Reports should be sent in on the first of the month following the month being reported. For example, May estimates should be sent in at the beginning of June.

The form is rather straightforward, with the consecutive report number and the number of sheets in the report appearing on the first line. The month for which the report is being made goes on the second line, and information about the observer and location is written on the next four lines. If you do not use UT, indicate what you do use on line five and information on your instrument on line six. The variables should be listed in the ascending order of their Harvard designation, followed by their letter designation, the JD, the decimal time, the magnitude estimate, the comparison stars used, and any remarks that you feel are necessary, such as "10 degrees above horizon" or "high clouds"—anything you think would have an influence on the listed estimate.

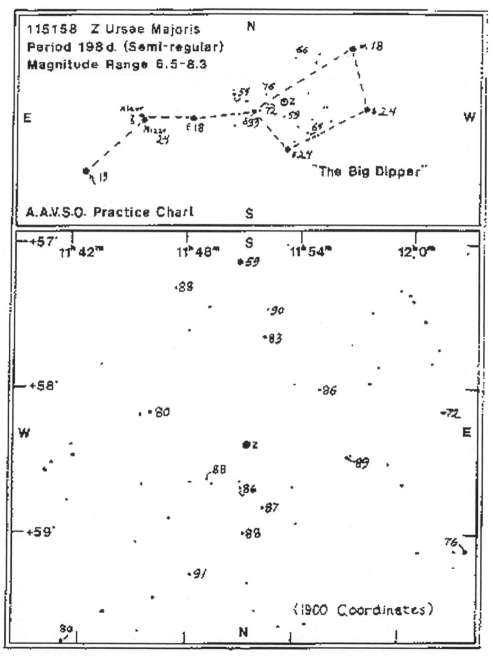

FIGURE 12-10 *AAVSO practice chart for variable stars. (Reprint by special permission of the American Association of Variable Star Observers, through its director, Janet Mattei)*

THE AMERICAN ASSOCIATION OF VARIABLE STAR OBSERVERS
25 Birch Street, Cambridge, MA 02138, USA

VARIABLE STAR OBSERVATIONS

AAVSO Observer Initials

Sheet _____ of _____   Report No _____

For Month of _____ Year _____

Observer _____

Street _____

City _____ State _____

Country _____ Zip Code _____

Time Used, GMAT or _____

Instrument(s) _____

For AAVSO HG Use Only

Received _____

Entered _____

Verified _____

Chart
P = AAVSO preliminary
S = AAVSO standard

Designation	Variable	Jul.Day+Dec.	Magn.	Key*	& Remarks	Comp. Stars	Chart/Scale/Date

Total Number Inner Sanctums*	Total Number Observations Reported

* KEY field contains AAVSO-selected one-letter abbreviations for REMARKS. See other side for list.
* An Inner Sanctum is an observation which is magnitude 13.8 or fainter, or <14.0 or fainter.
Observations should be sent to AAVSO Headquarters 25 Birch Street, Cambridge, MA 02138 USA, as soon as possible after the first of each month.

(turn sheet over)

FIGURE 12-11 *AAVSO monthly report chart (front). (Reprint by special permission of the American Association of Variable Star Observers, through its director, Janet Mattei)*

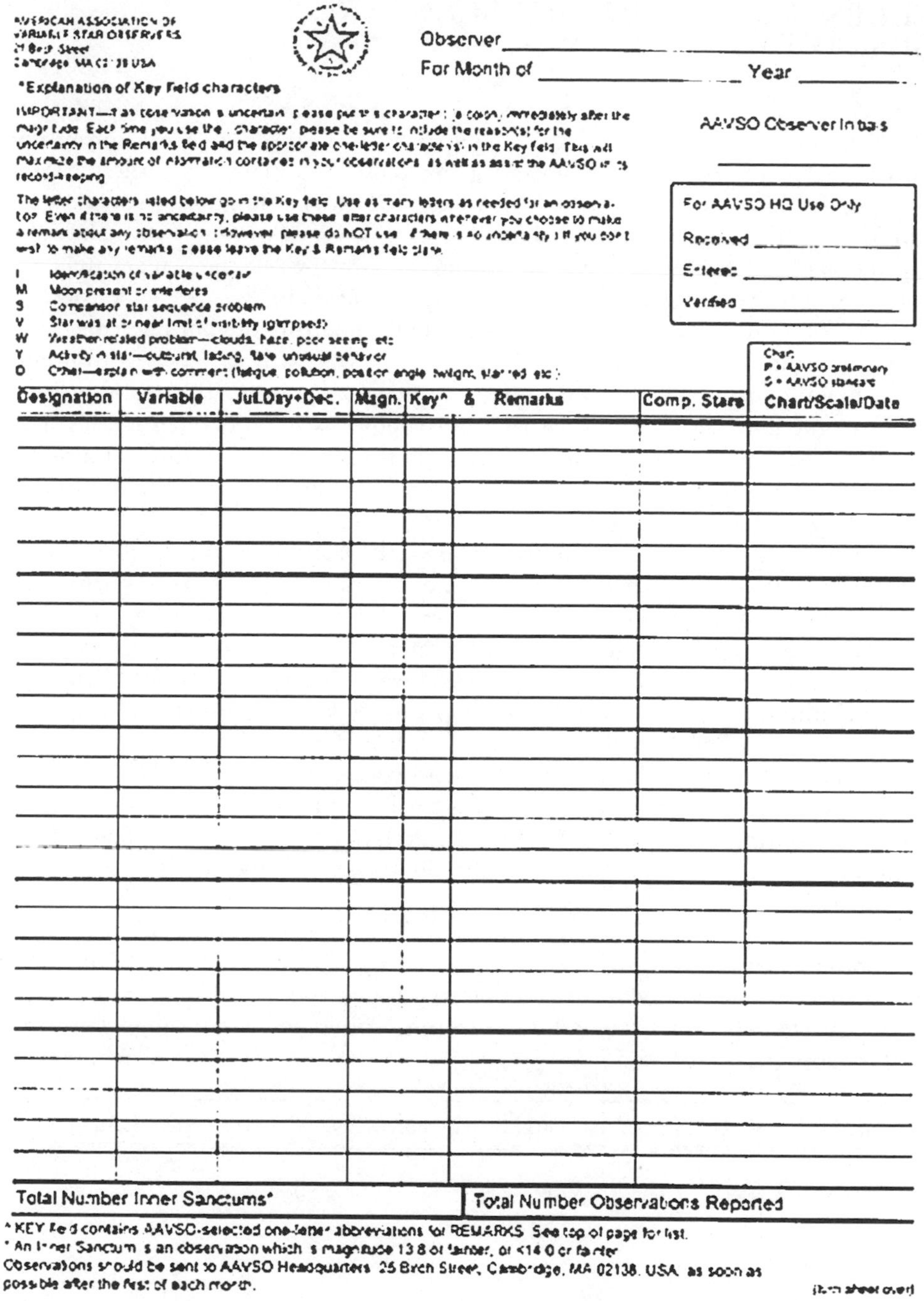

AMERICAN ASSOCIATION OF
VARIABLE STAR OBSERVERS
25 Birch Street
Cambridge, MA 02138 USA

Observer ______________________
For Month of ______________ Year ________

AAVSO Observer Initials

____

*Explanation of Key Field characters

IMPORTANT—If an observation is uncertain, please put this character : (a colon) immediately after the magnitude. Each time you use the : character, please be sure to include the reason(s) for the uncertainty in the Remarks field and the appropriate one-letter character(s) in the Key field. This will maximize the amount of information contained in your observations, as well as assist the AAVSO in its record-keeping.

The letter characters listed below go in the Key field. Use as many letters as needed for an observation. Even if there is no uncertainty, please use these letter characters whenever you choose to make a remark about any observation. (However, please do NOT use : if there is no uncertainty.) If you don't wish to make any remarks, please leave the Key & Remarks field blank.

For AAVSO HQ Use Only

Received ____________

Entered ____________

Verified ____________

I Identification of variable uncertain
M Moon present or interferes
S Comparison star sequence problem
V Star was at or near limit of visibility (glimpsed)
W Weather-related problem—clouds, haze, poor seeing, etc.
Y Activity in star—outburst, fading, flare, unusual behavior
O Other—explain with comment (fatigue, pollution, position angle, twilight, star red, etc.)

Chart:
P = AAVSO preliminary
S = AAVSO standard

Designation	Variable	Jul.Day+Dec.	Magn.	Key*	& Remarks	Comp. Stars	Chart/Scale/Date

Total Number Inner Sanctums° | Total Number Observations Reported

* KEY Field contains AAVSO-selected one-letter abbreviations for REMARKS. See top of page for list.
° An Inner Sanctum is an observation which is magnitude 13.8 or fainter, or <14.0 or fainter.
Observations should be sent to AAVSO Headquarters, 25 Birch Street, Cambridge, MA 02138, USA, as soon as possible after the first of each month.

(turn sheet over)

FIGURE 12-12 *AAVSO monthly report chart (back). (Reprint by special permission of the American Association of Variable Star Observers, through its director, Janet Mattei)*

TELESCOPES FOR VARIABLE STAR OBSERVING

Variable star observing is where amateur astronomers can really make their presence felt. You don't need a lot of expensive equipment. In fact, serious work can be done with a good pair of binoculars. If you are going to use binoculars, try to select variables that fade no fainter than magnitude 9. Appendix H lists a number of variables with minima brighter than magnitude 10 that are suitable for observation with binoculars. If you don't have a telescope but are planning to purchase one, binoculars are a good way to learn the techniques and the discipline you'll need once you get a telescope.

Your observing list of variables depends on the aperture of your telescope and its light-gathering power. It is a good idea for you to stay with variables with minima just below the limiting magnitude of your telescope's aperture. Table 12-2 lists various instruments and the faintest variable you should try to observe with each one. A number of things can determine your variable threshold: The seeing conditions of your observing site and your own visual sensitivity might increase or decrease the lowest magnitude you can easily see.

One thing that also enters into the picture is the fact that not being able to see the variable is, in itself, a valuable observation. For example, an 80-mm (3.1-inch) retractor has a limiting magnitude of 11.5. This should limit the variables on your list to those with minima of magnitude 11.2 or brighter.

When making your observations you will want to get the most out of your telescope's abilities. As with planetary observing, each type of instrument has its particular advantage. A short-focal-ratio system, such as f/4 or f/5, will give you wide, bright fields of view that make it easy to identify variable star fields. Telescopes with longer focal ratios, f/8 and up, will have a smaller but darker field of view, giving you better contrast between the stars and the dark background. The important thing in variable star observing is not the focal ratio or even the aperture of the telescope, but the person using the telescope.

Another good thing about variable star observing is that you don't need a bevy of eyepieces. For example, with an 80-mm telescope, a 26-mm plossl and an 18-mm orthoscopic would be fine. The 26-mm would give you a field just over 1 degree across—perfect to find a star field; the 18-mm orthoscopic would give you a good field to estimate.

TABLE 12-2. VARIOUS SIZE AMATEUR INSTRUMENTS AND VARIABLE STAR LIMITS

APERTURE (IN.)	LIMIT OF VARIABLE	APERTURE (IN.)	LIMIT OF VARIABLE
Eye	6.2	10.0	14.5
1.4	9.5	12.5	15.4
2.0	10.3	13.1	15.5
2.4	10.7	14.0	15.6
3.1	11.3	16.0	16.0
4.0	11.8	17.5	16.1
6.0	13.0	20.0	16.4
8.0	13.6		

Variable star observing will teach you the limitations of your instrument, be it large or small. It will show you how to stay within those limitations while still contributing to science. Telescopes and eyepieces aside, you do need a few things. The first is the discipline to go out and make your estimates at every opportunity. If you join AAVSO, you will be issued charts on two or three variables to get you started. And on the group's Web site, you can read more about how to search for variables—and even the trials and tribulations of other members.

You will also receive a list of predictions for the stars in AAVSO's observing program. This handy little publication tells you the approximate dates predicted by the AAVSO computer for the star's maximum and minimum. It also tells you, when possible, whether the variable is brighter than magnitude 11 or, if the minimum falls this low, fainter than 13.5. With this list you can plan your sessions at the telescope and expand your own list of variables as rapidly or slowly as you like. AAVSO recommends that you observe each variable on your list twice a month.

And how else can a variable star watcher contribute to science? According to the AAVSO, professional astronomers have relied on amateur variable star watchers for real-time information or same-time observations—thus the success of hundreds of observing programs, particularly those using satellites. These include the Apollo-Soyuz mission, the International Ultraviolet Explorer (IUE), the Roentgen satellite (ROSAT), the High Precision Parallax Collecting Satellite (HIPPARCOS), the Extreme Ultraviolet Explorer (EUVE)—and even the Hubble Space Telescope (HST). In addition, a significant number of rare events have been observed with these satellites, thanks to the timely notification by the AAVSO and its members.

13

OBSERVING DEEP-SKY OBJECTS

We are . . . approaching more amazing regions and fresh scenes will open upon us of inexpressible and awful grandeur.

REV. T. W. WEBB

If the solar system offers the observer an abundance of objects to study, the deep sky offers a veritable plethora of objects for the amateur's telescope. The star clusters, nebulae, and galaxies that lie within reach of even the small telescope can keep the amateur at the eyepiece for a very long time.

Deep-sky objects (DSOs) fall into three basic categories: *Star clusters* are groups of stars bound by gravity and that move together through space. A *nebula* is a filmy sheet of glowing gas surrounding a star or star cluster. *Galaxies* are "continents" of stars, millions and even billions of light years away from us, floating through the universe.

Countless amateurs have devoted their observational time to studying these objects because of the staggering beauty of their structure. They have even been able to make a few contributions to science along the way.

TELESCOPES, IMAGE BRIGHTNESS, AND THE DEEP SKY

The observer of solar system objects is looking at the bright image of a planet spread over an extended area. The double star or variable star observer is looking at point images of individual stars. But the types of objects that interest deep-sky observers are quite different. DSOs are usually faint conglomerations of stars, gas, and dust spread over a relatively large area of the sky. They are classified as extended objects like planets—but their brightness is hundreds and even thousands of times fainter.

With any object you see, whether with the naked eye, binoculars, or a telescope, a limited amount of light reaches your eye. Because deep-sky objects cover a large area, the light available to your eye is spread evenly over an equally large area. As a result, you see a faint object.

When you view a DSO through a telescope, you are forcing the available light to cover a larger area. Because the image is magnified, its brightness decreases. If you continue to increase the magnification of the image, you will make things worse. There is a magnification point at which there is too little available light—thus the image blends into the background and disappears.

The key to successful deep-sky observing is to get the brightest image possible from these faint objects. The observer measures the brightness of each deep-sky object three ways: by its integrated magnitude, by its normal brightness, and by its telescopic brightness. The *integrated magnitude* assigned to DSOs is not quite the same as the magnitudes given for point-source objects like stars. Because a DSO is an extended object, the integrated magnitude is the magnitude the object would have if all of its light were compacted, or integrated, into a single point source. As the size of the area that the object covers increases, it will become harder to see the object— even when a relatively high magnitude figure is listed for it.

The normal brightness of a DSO is the brightness of the object as viewed with the naked eye. But a magnitude-9 galaxy is not visible to the naked eye, so how is it possible to measure its normal brightness? The objective of a telescope or binoculars is an extension of the human eye and increases the amount of light available based on the objective's light-gathering power. As you increase the size of the objective, more light passes to the eye, allowing dimmer objects to be seen. When you use an eyepiece with an exit pupil (remember them?) that equals the size of the eye's pupil, your eye is able to use all the light the objective collects. This normal exit pupil allows the eye to sense this image across the entire retina—so the telescopic brightness seen is the same as the object's normal brightness.

All this relates to the type of telescope best suited for observing these "faint fuzzies," as they are affectionately known. To achieve an exit pupil of 7 mm with an f/11 telescope, you would need an eyepiece of almost 80 mm focal length—a great deal of glass. But with a telescope of f/4 you only need a 28-mm focal length eyepiece to get a 7-mm exit pupil. While 80-mm eyepieces are available, they are hard to come by and very expensive. A 28-mm eyepiece, on the other hand, is easily obtained and won't put your budget on the critical list. So the first requirement for a deep-sky instrument is that it have a short focal ratio, usually f/4 or f/5. The extremely short-focal-ratio instruments, f/4 and shorter, are called *rich-field telescopes* (RFTs).

The normal exit pupil also sets a limit on the object's maximum angular size, giving it enough contrast to stand out from the background field of view (see Table 13-1). Objects that are magnified to a larger size than this limit begin to lose contrast and grow dimmer, until they fade into the background and disappear. Objects smaller than this limit will appear as bright patches in the field of view. The larger the aperture of the telescope, the more light that is gathered and sent to the eye—and the smaller this size limit becomes. This way, a smaller DSO can be observed at maximum brightness. So the next requirement for a deep-sky telescope is to have a large aperture. Combining a large aperture and a short focal ratio in a telescope allows you

TABLE 13-1.   CONTRAST LIMITS FOR INSTRUMENTS OF VARIOUS SIZES

APERTURE (MM)	(IN.)	SMALLEST ANGULAR DIAMETER
35	1.4	1.0°
50	2.0	43.0′
60	2.4	33.0′
80	3.1	27.0′
100	4.0	21.0′
150	6.0	14.0′
200	8.0	11.0′
250	10.0	8.4′
317.5	12.5	6.8′
330	13.1	6.5′
355	14.0	6.1′
405	16.0	5.3′
444	17.5	5.0′
500	20.0	4.3′

to use a higher magnification on a deep-sky object and still maintain adequate contrast between the image and its background.

The aperture of the deep-sky observer's telescope is a matter of individual choice. Some observers are quite happy with a 4- or 6-inch RFT, while others crave the dimmer limiting magnitude and greater image contrast of a 13- or even 17-inch aperture. With the introduction of thin-mirror technology and compact dobsonian mounts, many amateurs are opting for larger-sized instruments. Apertures in the 20-inch range are considered portable and easily handled by one person.

AIDS FOR THE DEEP-SKY OBSERVER

Deep-sky observers use a visual trick to help them see the faint fuzzies they chase through the night sky. *Averted vision* is an excellent way to improve your ability to see deep-sky objects. When you look directly at an object, most of the light falls on the center of the retina. This area, the *fovea centralis*, is composed primarily of cones that react only to bright light. When the light from a DSO falls on the fovea, you will see only a dim image.

To use the more numerous rods sensitive to faint light, you must direct the image away from the fovea and onto the rods. To do so, simply look at the side of the field of view. When the image falls on the rods, it will become brighter and you will be able to see it much better. Using averted vision is much the same as using peripheral vision to detect something out of the corner of your eye. When using averted vision you look at one part of the field of view while directing your attention to another part. Getting used to using averted vision takes a little practice, but after awhile it becomes almost an automatic response to detecting faint objects.

If you are going to get serious about deep-sky observing, you are going to need a good star atlas. Norton's is great for the beginning observer, but something with more detail is needed in order to hunt down really faint DSOs. It is easier to find a faint galaxy in a field crowded with similar objects—if you can pick out and identify an individual star, from which you can get your bearings. To do this, you must have an atlas with a deep limiting magnitude.

A good rule of thumb is to use a chart and a finder with the same limiting magnitude. This means an atlas with at least a limiting magnitude of 8. Norton's only goes down to magnitude 6, but Wil Tirion and Roger W. Sinnott's excellent *Sky Atlas 2000* goes to magnitude 8. The recently published *Uranometria 2000* lists stars down to magnitude 9.5 and is often called *the* atlas for the deep-sky observer. And there are the more advanced atlases, including the *Millennium Star Atlas* by Roger W. Sinnott and Michael A. C. Perryman, the-sky atlas based on the European Space Agency's (ESA's) Hipparcos satellite.

An atlas designed specifically for the deep-sky observer is the *Sarna Deep-Sky Atlas*. Deep-sky objects of various types are listed on 102 star charts that have a scale of 1 mm equal to 1 arcsecond. Compared to the scales of Norton's atlas, *Sky Atlas 2000*, and *Uranometria*, this is the largest-scale atlas available. The *Sarna Deep-Sky Atlas* is unique because it not only lists the right ascension and declination of its 254 objects, it also gives the observer simple directions for star hopping to the object using bright stars and easy-to-find asterisms.

Another crucial item for deep-sky observers is the finder. Because you are looking for very faint objects, you need to be able to pick your way across the sky to your target. Star hopping between bright stars and asterisms is an excellent way to do this—but you will need an instrument with a limiting magnitude low enough to allow you to find your key asterisms on your way to that faint fuzzy. The majority of finders available today are either 6 × 30 (objective 30 mm, magnification 6×) or 8 × 50 (objective 50 mm, magnification 8×); a few companies even market finders with an 80-mm objective. For most amateurs, unless you are going the 20-inch dobsonian route, a 6 × 30 or 8 × 50 finder is more than adequate.

Many finders sold today come with a right-angle star diagonal that eliminates the need to bend over in order to look through it. While this may be convenient, it can make things look pretty weird in the finder's field. Remember the image orientation of the standard astronomical telescope? A finder is subject to the same orientation—after all, it is an astronomical telescope even though it doesn't boast a huge aperture. Putting a star diagonal on the finder is only going to confuse you by giving you an inverted and reversed image that will not match any of your star charts.

When you buy a finder telescope with a right-angle finder, check its image orientation. If the field is reversed and inverted, most manufacturers make an adapter that eliminates the right-angle feature, allowing you to sight straight through the finder. If you want the right-angle feature, switch the inverted/reversed diagonal with an amici diagonal so that you will have the proper orientation in your finder field.

Deep-sky observing requires a few things from the sky that other areas of observation do not need. If you are observing bright objects such as the Moon, planets, bright double stars, and even the brighter variables, you will be relatively unaffected by light pollution. The observer of the deep sky, however, needs all that extra light like he needs a wire brush to clean his mirror. The farther away one can get from the city lights, the better it is for the deep-sky observer.

This aversion to light also means that the deep-sky observer wants to avoid that old nemesis, the Moon. Deep-sky folk have little use for our natural satellite and plan their observing schedules to coincide with the new Moon whenever possible. Deep-sky observation also calls for the atmosphere to be exceptionally clear. When the stars stand out like sparkling diamonds against a velvet background—and you can see easily stars of magnitude 6—it is a perfect night for observing the deep sky. Because objects like nebulae and galaxies are faint and extended, they show up best during conditions of high transparency.

For the deep-sky observer, deep sky means dark sky. Where can the deep-sky observer go? There are no deep-sky parks listed on any maps. The deep-sky observer must often bundle the telescope into the car and hit the road. Depending on the size of the city you are fleeing, a drive of between 40 and 100 miles should bring you to nice dark skies. If you are lucky, you might only have to drive for an hour or two to reach a dark site. You can make the round trip, observe, and be back home for breakfast.

Many amateurs make dark-sky weekends a family affair, bringing the entire clan along for a weekend of camping and observing in a nearby state park. If your family is not the outdoor type, you may be able to find a nearby motel that can act as a base camp. Some amateurs have found remote, quiet nooks off the busy interstates and return there time after time.

If the site you find is on private property, please get permission before you set up your gear. There is nothing like hearing the low growl of a farmer's dog to start you and your telescope shaking. If you take the time to explain to the property owners what you are doing and offer them a look through your telescope, you will usually find them only too happy to have you around.

For the amateur stuck under the light-polluted skies of the city, there are ways to overcome this handicap. A number of companies are producing a line of filters especially designed to cut through light pollution and bring the deep sky to the city observer. *Light pollution reduction (LPR) filters* screen out the light from some types of sodium vapor lamps, minimizing skyglow. These deep-sky filters can improve city observing by filtering out light pollution and increasing the contrast of deep-sky objects.

Other deep-sky filters select particular wavelengths of light for transmission to the eye. Ultrahigh-contrast (UHC) filters increase the overall contrast between faint planetary nebulae and the background sky. Oxygen-III (O-III) filters select wavelengths of light that complement the UHC filter by allowing through wavelengths unique to planetary nebulae. Hydrogen-beta line filters (H-BETA) are sensitive to light common in reflection- and emission-type nebulae.

Because deep-sky telescopes are usually more portable than others, it's important to have a source of power for the drive unit, dew zappers, and other little electrical goodies that you may use. It's not advisable to use an adapter that lets you tap into your car battery; many astronomers have run down their car batteries, something that is difficult to remedy so late at night. A good alternative is a marine battery, which offers an excellent, portable source of power for your drive and electrical accessories.

No matter where or what you observe, you are going to be faced with one adversary that is even more persistent than light pollution. It will ruin your nights by spreading itself over your optics, your charts, and your glasses. It comes on perfectly clear, transparent evenings and nights when the seeing for planetary work is better than excellent. And it's about every amateur's "friend": dew.

Dew might look pretty on the filmy strands of a spider's web or the petals of a flower, but it doesn't look good on glass. Dew on your optics can cut off the light entering your telescope just as surely as putting a bag over the lens. The light hitting the dew-covered objective scatters every which way, turning the telescope into a very expensive kaleidoscope.

Over the years, amateurs have come up with a wide variety of ways to defeat this enemy. Some use fans to circulate the air over the objective. Some use hair dryers to dry off the offending dew. Some, admitting defeat, simply pack up and head for home. The best way to treat dew on your optics is to be prepared for it.

The easiest way to combat dew is to have an adequate dew cap at the end of your tube. For retractors or catadioptrics, the length of this dew cap should be equivalent to the diameter of the primary. So for a 4-inch retractor the dew cap should be 4 inches long. With reflectors, the tube length should extend at least 4 inches past the secondary support, preferably more. One way is to place *heat ropes*, insulated strips of nichrome wire, around the optical surfaces of the telescope. Running a very small current through the ropes heats them up, transferring their heat to the optical surface. This slight increase in heat is all that is needed to prevent dew from forming.

On a retractor or catadioptric, running a heat rope around the inside of the dew cap is recommended. With reflectors, a heat rope on the secondary mirror as well as along the inside of the tube near the primary mirror is recommended. You also can protect your finder from dew using the same technique.

Some amateurs have even devised heaters for their eyepieces—although the eyepiece is the only element of the optical system that can be removed and warmed by itself. If dew forms on the eyepiece, you can hold it under your car's heater to dry it off or use your own body heat to get the same result. Whatever you do to combat dew, if it forms and you can't heat it away—*do not* wipe it off the optics. This could scratch the optics, and although dew will eventually dissipate, scratches never go away.

Dew also will soak your charts and notebooks. Star atlases that come in individual sheets, such as *Sky Atlas 2000* and the *Sarna Deep-Sky Atlas*, can be laminated in plastic. This won't work however, with book-type atlases like Norton's and *Uranometria*. The easiest way to protect these atlases is to cover them with a sheet of stiff, transparent plastic; you can see the chart directly through the covering.

LETTERS, NUMBERS, AND LISTS

When you are just getting into deep-sky observing, it helps to have a list of easy-to-see objects. Probably the best list of deep-sky objects ever compiled was put together by a French comet hunter who had very little use for these objects because they were so easily confused with comets. Charles Messier began compiling his list in 1758, when he came across the Crab nebula while looking for a comet he had found two weeks earlier. The object he saw puzzled him because it "was a little like . . . the comet I had observed before, however, it was too bright."

To save himself and other comet hunters the trouble of misidentifying this "whitish light," Messier recorded its location on one of his charts. The list that Messier began that night was originally published in 1774 and included 45 deep-sky objects. In 1780 Messier published a supplement to his original list that included 23 more objects. In 1781 he added 35 new objects, 24 of which were discovered by a colleague named Pierre Mechain, and published the final version of his catalog containing 103 objects. In the 1920s, the French astronomer Camille Flammarion and American astronomer Dr. Helen S. Hogg added another seven deep-sky objects, the descriptions of which they had found in Messier's and Mechain's notes.

The *Messier Catalogue* includes a total of 110 objects. All of these objects are easily visible through amateur instruments. Known by the preface M, followed by their catalog number, these objects comprise the *créme de la créme* of deep-sky objects.

An amateur interested in deep-sky observing would be well served to undertake a program to find and record all 110 Messier objects. Such a project should be pursued at a comfortable pace. There is no time limit unless you choose to impose one. There is no sense in rushing through such a pleasing task. Each object should be observed with an eye toward the detail you can make out with your instrument. Keep your observations in your observing log, including a written description and sketch of each object. "Doing the Messiers" will further develop your observational skills. If you use the starhopping technique to locate these objects, you will come away from the project with an even better knowledge of the night sky and the constellations.

Once your appetite for the deep sky is whetted by the *Messier Catalogue*, where can you turn? During the late nineteenth century, J. L. E. Dreyer cataloged deep-sky objects down to magnitude 12 for almost the entire sky. The *New General Catalogue of Nebulae and Clusters of Stars* (NGC) lists over 8,000 deep-sky objects for the amateur to chase. In the 1970s astronomers from the University of Arizona updated the information available on these objects and published the *Revised New General Catalogue of Nonstellar Astronomical Objects* (RNGC). The RNGC includes the original object descriptions from Dreyer's NGC, information from the National Geographic Society and Palomar Observatory Sky Survey conducted in the 1950s and 1960s, and updated position information.

After a survey of all these objects has been completed, the next logical target for the deep-sky observer is to observe as many of the NGC objects as his or her telescope can reach. Twenty years ago, it would have been unthinkable for an amateur to even consider a study of the complete NGC list. But today's large-aperture telescopes put such a study well within the reach of the dedicated deep-sky observer.

STAR CLUSTERS

Some of the most spectacular deep-sky targets for the amateur's telescope are star clusters (Figs. 13-1 and 13-2). Some, like the Pleiades (M45) and the Great Hercules cluster (M13), are visible through binoculars; others require a telescope.

Star clusters fall into two basic categories. *Open clusters* are groups of stars bound by gravity that move through space together. They can range from loose

FIGURE 13-1 *M55, a large globular cluster about one quarter the angular size of the Moon. It is about 20,000 light years away and may have a luminosity greater than 100,000 times that of the Sun. (NASA)*

conglomerations of stars, such as the Beehive cluster (M44) in Cancer, to tightly packed groups containing hundreds of stars, such as the Double Cluster in Perseus. The biggest problem when trying to observe some of these clusters is identifying them. Although bright clusters may be immediately apparent, fainter ones can blend into the rich background of stars.

If you have access to some of the photographic all-sky atlases such as the *Palomar Sky Survey* or the *Atlas Stellarum*, this becomes less of a problem. Most of the clusters listed in the NGC can be picked out easily because these atlases have very deep limiting magnitudes. All you have to do is compare your field drawings with the atlas or trace the field from the atlas for comparison at the telescope. These atlases are very expensive, however, and out of the price range of the beginner. Vehrenberg's *Atlas Stellarum* is a two-volume set of charts that costs over $200; you really don't want to know the cost of the *Palomar Sky Survey* plates.

An easier and much less expensive way to get a peek at what your target looks like is to get The Webb Society's *Deep Sky Observer*, published quarterly. They also offer handbook volumes that are indispensable references for anyone interested in deep-sky observing—from galaxies to double stars. For example, the third volume, *Open and Globular Clusters*, lists descriptions and drawings of 205 of the brightest open clusters in the sky. With this handbook, you can compare the drawing of the cluster directly with what you see at the eyepiece.

FIGURE 13-2 *M8 in Sagittarius is a globular cluster. (R. M. Sandy and NASA)*

Open clusters are classified under a system devised by Robert Trumpler of the Lick Observatory in the late 1920s. His system rated open clusters according to three characteristics. The *concentration* of the open cluster can be

  I Detached from the surrounding star field with a strong concentration of stars toward the cluster's center.
 II Detached from the surrounding star field with a weak concentration of stars toward its center.
III Detached from the surrounding star field with no apparent concentration of stars toward the cluster's center.
IV Not well detached from the surrounding star field.

The second characteristic assigned by Trumpler concerns the *range of brightness* for the open cluster:

 1 Cluster has a small range of brightness between its brightest and faintest member star.

2 Cluster has a moderate range of brightness between its brightest and faintest member star.
3 Cluster has a large range of brightness between its brightest and faintest member star.

The final characteristic is the *richness* of the cluster measured in terms of the number of stars composing the cluster:

p   Indicates a poor cluster, having less than 50 stars.
m   Indicates a moderately rich cluster, with between 50 and 100 member stars.
r   Indicates a rich cluster, with over 100 stars.

A cluster like the Pleiades has the classification of I-3-r, indicating that it appears detached from the surrounding star field, has a large range of brightness from the faintest to brightest members, and has over 100 stars associated with it.

The second type of star cluster is the *globular cluster*. Globular clusters are compact groups of 100,000 or more stars squeezed into a distinctive spherical shape. Because of the vast numbers of stars they contain, these objects appear significantly brighter than open clusters and are much easier to pick out from the stellar background.

Like open clusters, globular clusters are classified using a scale denoting their concentration. This scale was devised by Harlow Shapley of the Harvard College Observatory in the 1920s. The classification of globular clusters is much simpler than that for open clusters. It ranges from 1 for the most highly concentrated to 12 for the least concentrated. The Great Hercules cluster (M13) rates a 5 on Shapley's scale, telling the observer that it is of about average concentration. The thing to remember when using this scale is that the smaller the number, the higher the concentration of stars in the object.

STAR CLUSTERS FOR THE TELESCOPE

Probably the best-known open cluster is the Pleiades (M45). It has been known and recognized as a related group of stars since the days of Homer. Easily found with the naked eye in the constellation Taurus, the cluster is best seen with a wide-field, low-power eyepiece or with binoculars. To the naked eye, from 6 to 12 stars have been reported visible. With a small telescope, close to 80 stars are visible.

During the last half of the nineteenth century, a debate raged concerning the visibility of a nebula surrounding the Pleiades member Merope. Observers said that this nebulosity was easily visible in a small instrument, but when a larger one was turned on the object, it was invisible. One observer said that the nebula was visible with his 2-inch finder, but invisible with the 11-inch main telescope. S. W. Barnard, using the 18-inch Alvan Clark retractor at Dearborn Observatory, could not see the nebula. Thus it did not exist. It is heartening to know that some things in the universe are visible to the amateur with a small instrument—and denied to the owners of larger telescopes.

Another naked-eye open cluster is found in the constellation Cancer. The Beehive cluster (M44) was seen by Galileo to contain "a mass of more than 40 small stars." Because the cluster is so large, almost 2 arcminutes across, it is best seen in wide-field instruments such as a rich-field telescope or even binoculars. The Beehive has a Trumpler classification of II-2-m.

A final naked-eye cluster is really a pair of open clusters aptly named the Double Cluster. Easily found midway between the *W* of Cassiopeia and the main body of Perseus, they appear as a bright, hazy spot against the sky. For some reason these beauty clusters were overlooked by Messier and are not included in his catalog. Officially known as NGC 884 and NGC 869, these objects make up one of the most beautiful sights for the small telescope.

For the best view, use an eyepiece with a field of at least one degree. The Double Cluster fills the field of view with a diamondlike sprinkling of stars. From their obvious intensity, the Trumpler classification of I-3-r for each cluster, the same as for the Pleiades, is well deserved. While you are drinking in the beauty of the clusters, take a few moments and see if you can detect colored stars in either one. There are a number of red supergiants in and around the clusters. The eastern cluster (NGC 884) has three red stars visible inside it, and there is a very prominent red star between the two clusters.

Our next open cluster, M37, is only visible through a telescope. Located in the constellation Auriga at RA 5h52m, Dec +32°33′, M37 is the brightest and best of the three Messier clusters in the constellation. The biggest difference among the three Auriga clusters—M36, M37, and M38—is the brightness range of the stars in each cluster. All three are moderately rich to rich. Two, M36 and M37, appear to have a weak concentration of stars toward their centers; M38 has no apparent concentration. M37 appears the brightest because it has the smallest brightness range of the three combined with the richest membership—or more stars with less difference in their brightness.

M29 (Fig. 13-3), in the constellation Cygnus (at RA 20h24m, Dec + 38°32′), on the other hand, is a difficult object to pick out of the background stars. This is because it has no apparent concentration of stars at its center (III), has a large brightness range (3), and is a star-poor cluster (p). It's definitely easy to overlook.

NGC 2362 (Fig. 13-4)—located in Canis Major at RA 7h18m, Dec 24°57′ and centered near the magnitude-4 star Tau—is an example of what a strong concentration of a few stars can do to the overall appearance of a cluster. The cluster has a Trumpler classification of I-3-p. You would expect that; it has a large brightness range and relatively few stars, so it appears as a dull little cluster. But with most of the stars concentrated toward the center of the cluster, it appears "small but densely packed with faint stars" in a 4-inch telescope.

Probably the best-known globular cluster is M13, the Great Hercules cluster. We have already had a peek at this with our binoculars, so let's compare it with our view through the telescope. With an integrated magnitude of 5.8, M13 is just visible to the naked eye when the skies are extremely transparent. The cluster was discovered in 1774 on a chance observation by Edmund Halley of comet Halley fame. Messier observed it a short time later and declared emphatically that the nebula contained no stars. This object can be completely resolved by a 6-inch telescope at 130×—thus we know Messier was wrong about it not containing any stars.

In 10 × 50 binoculars the cluster is visible only as a bright smudge against the background stars. Through a small telescope, it begins to take on the classic characteristics of a globular cluster, with a milky center surrounded by glittering specks of

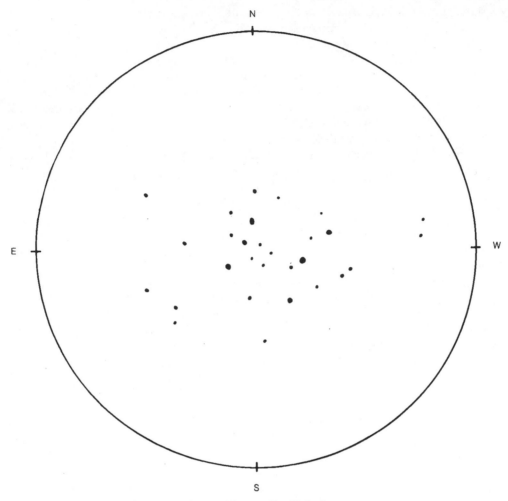

FIGURE 13-3 *M29, an open cluster in Cygnus. (Courtesy Dan Troiani)*

starlight. Globulars take magnification fairly well, so don't be afraid to use a higher power. For example, at 36×, an 80-mm shows a round shape with a very bright center; with a 100×, you can resolve some of the outlying stars, but nothing near the center.

Located north of M13 in the same constellation—at RA 17h17m, Dec +43°08′— is the globular M92. With a Shapley classification of 4, M92 is slightly more concentrated than M13. Using a 4-inch SCT, Barbara Lux of McKeesport, Pennsylvania, describes it as "bright and beautiful with its central condensation and diffused edges plainly seen."

Slightly less concentrated than M13 is M3 in the constellation Canes Venatici, at RA 13h42m, Dec +28°23′. With a Shapley classification of 6, M3 is seen as a looser cluster than either M13 or M92. A 6-inch telescope will resolve the stars around the edge of the cluster, but nothing toward the center of the globular.

Our final globular is the loosest of all we have seen: M4 can be easily found near the star Antares in the constellation Scorpio. With a Shapley rating of 9, it is the only

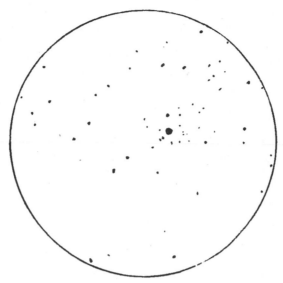

FIGURE 13-4 *NGC 2362, an open cluster in Canis Major. (Courtesy Suresh Sreenivasan)*

globular we've observed that can be at least partially resolved with a small telescope. A 3- or 4-inch telescope easily shows the outer edges broken into individual stars; a 4-inch will partially resolve the center of the cluster.

A number of other clusters suitable for observation with a small telescope are listed in Appendix F.

You might have noticed that many of the open-star clusters are observed in the autumn or winter sky, whereas the globulars are found in the spring and summer. There is a simple reason for this—and it relates to how the clusters are located throughout the galaxy. Open clusters form in the spiral arms of the Milky Way galaxy; globular clusters tend to be around the center of the galaxy. Many (but not all) globulars are found in a halo, orbiting like satellites around the center of the galaxy. When we face outward away from the center of the galaxy, as we do during the winter months, we are looking through the arms of the galaxy—thus we see many more open clusters. In the summer months, when the Earth's orbit allows us to look toward the center of the galaxy, we see more globular clusters.

NEBULAE

At one time, the term *nebula* referred to a number of faint, apparently glowing objects. This included galaxies, which are completely different as we will see. Today, astronomers break nebulae into three major classes:

Dark nebulae are lanes of dust thick enough to block the light from the stars behind them. They appear as dark lanes seemingly void of stars and can be found along our line of sight toward the center of the galaxy. A dark nebula can block light coming through it by as much as three magnitudes. The Great Rift, running from Cygnus to Sagittarius, is a good example of a dark nebula. The Milky Way seems to be split in two as it runs south out of Cygnus, through Aquila, and into Sagittarius, where it recombines into one large star field. The Great Rift is caused by dust lying between us and the center of the galaxy—the direction in which we look when facing those constellations. Dark nebulae are best observed with nothing more than the naked eye.

Bright nebulae are knots of gas and dust that are illuminated by stars buried within or around the nebula. Bright nebulae can be illuminated from behind, forming *emission nebulae*, type E, which shine from the light of stars embedded in them, and whose light passes through them. Bright nebulae also can shine because of light from nearby stars reflecting off the dust making up the nebula, forming a *reflection nebula*, type R. A bright nebula also can be made by a combination of these two light sources, forming an *emission-reflection nebula*, type E + R.

Planetary nebulae are the shells of stars blown off into space as the star ages. The expanding shell of matter is illuminated by the light of the original star. This star is most often found near the center of the nebula, but is usually too faint to be seen in most amateur telescopes. The similarity in shape to a planet and the nebula's regular outline led Sir William Herschel and other eighteenth century observers to call the nebulae "planetlike"— and the name stuck.

Planetary nebula range from starlike, stellar points of light to rather irregular forms. Small planetaries, particularly the stellar types, can be difficult to pick out from the background stars. You can help identify these planetaries by using a diffraction grating. Hold it between your eye and the eyepiece and move it quickly back and forth. Because stars shine by self-produced light, their images will turn into tiny streaks showing their spectra. But because the planetary nebula shines from second-hand light from its star, it will not streak and appears as a tiny point of light. When this technique is used, the planetary nebula will appear the same with and without the grating.

NEBULAE FOR THE TELESCOPE

The best-known of all bright nebulae is found in the constellation Orion embedded near the tip of that figure's sword. M42 (Fig. 13-5), the Orion nebula, is the classic example of the combination emission-reflection nebula. Observers get a bonus when observing this object. The nebula is part of a vast complex of nebula and stars. We see the bright emission-reflection nebula and a dark nebula in the complex, called the Fish's Mouth, composed of dust and gas lying between us and the Orion nebula. Near this dark nebula, embedded in the wisps of gas, is a group of stars known as the Trapezium, which have burned their way through the bright obscuring dust making up the E + R nebula.

Although the Orion nebula is considered by many to be the most spectacular object on Messier's list, it is interesting to note that Galileo never turned his instrument on it. The discovery of the Orion nebula had to wait 46 years until 1656, when Christian Huygens turned his telescope to the stars in Orion's sword and made a double discovery: First he found the Great Orion nebula and three stars close together. A fourth star, first seen by Hooke in 1664 but not recorded, was rediscovered by Struve in 1826. These stars in the Trapezium and the Orion nebula are linked together—for the nebula is a stellar nursery, the birthplace of the stars.

FIGURE 13-5 *M42 in Orion. (R. M. Sandy and NASA)*

Also located in Orion is a fine example of a reflection nebula, M78. Located above the belt of Orion—at RA 5h47m, Dec +0°03'—M78 is easily seen even under moderately dark skies. Using an 8-inch SCT from his home just south of Chicago, Suresh Sreenivasan describes it as being a "light wispy nebulosity with two bright clumps of nebulosity embedded in it." (Fig. 13-6)

One of the easiest planetary nebulae for the small telescope is M57, the Ring nebula. It was first observed by Messier early in 1779. He described it as looking like a "patch of light, rounded but unresolvable" when viewed by the best telescopes of the time. Decades later, Sir John Herschel said it looked like a piece of gauze stretched over a hoop.

For most small instruments, this is an easy object, despite its lack of surface brightness and low apparent magnitude (9). This is because the Ring nebula sits almost alone in its field, with no real distractions to divert the eye's attention. With an 80-mm retractor the Ring nebula is easily found, even against the light-dappled skies over a big city. A dim, almost grayish-looking object, it takes magnification well.

Located in the constellation Gemini—at RA 7h29m, Dec +20°55'—is a real challenge to your growing observation skills: NGC 2392, the Eskimo nebula (Fig. 13-7). It

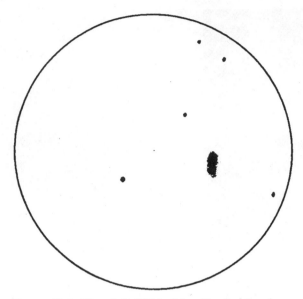

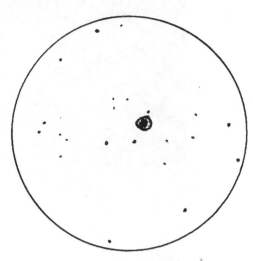

FIGURE 13-6 *The nebula M78 in Orion. (Courtesy Suresh Sreenivasan)*

FIGURE 13-7 *NGC 2392, the Eskimo nebula, is a planetary nebula in the constellation Gemini. (Courtesy Suresh Sreenivasan)*

challenges the owner of a larger instrument; the challenge is to pick out the details that gave the nebula its name. You will know this planetary when you find it by its blue-green color. Try using the grating technique described earlier to help you along. If you are using a scope less than 6 inches in aperture, this faint, magnitude-9.9 object may reveal the outer areas of its shell, but you will have to use averted vision. The 8-inch telescope will begin to reveal the mottling around the edges of the object, which makes it look like a tiny face peeking out from under a parka.

Other nebulae suitable for observation with a small telescope are listed in Appendix F.

GALAXIES

Galaxies are probably the most challenging of all deep-sky objects (Fig. 13-8). The vast numbers and wide diversity of these "island universes" inhabiting the night sky make them common targets for deep-sky telescopes.

For many of these objects, however, the telescopic requirements are a bit stiffer than what you have experienced so far. Most galaxies are not only faint, they have very low surface brightnesses, which makes them impossible to resolve in most amateur instruments. But galaxy hunting is a fun side road for the amateur to venture to from time to time. If you have a telescope in the 8- to 10-inch range you might even make a significant contribution to science by discovering a supernova on the edge of a faraway galaxy.

The key to observing a galaxy is the orientation it takes toward Earth. If the galaxy is almost face on, like M83, it has better contrast with the background and is

FIGURE 13-8 *The Whirlpool galaxy. (Courtesy of Lick Observatory)*

easily seen. A galaxy turned on edge, like NGC 4565, offers poorer contrast and is more difficult to see.

GALAXIES FOR THE TELESCOPE

At 2.2 million light years, the Andromeda galaxy, M31, (Fig. 13-9) is at the very limit of the universe seen with the unaided eye. Many objects are farther away, but none appear as majestic when viewed from our little corner of the universe. To the unaid-

FIGURE 13-9 *The Andromeda galaxy. (Courtesy of Lick Observatory)*

ed eye, the Andromeda galaxy appears to be a wisp of smoke or cloud lost among the stars. Over 1,000 years ago, the Persian astronomer Al-Sufi described this galaxy as "a little cloud" in his *Book of the Fixed Stars*.

The first person to look at the galaxy with a telescope was Simon Marius. A contemporary of Galileo, Marius turned his simple telescope on the Great Andromeda nebula in December 1612. He concurred with the cloudlike nature of the object, but noted that it "consisted of three rays; whitish, irregular and faint; brighter towards the center." Edmund Halley, second Astronomer Royal, also studied it and said it appeared to emit a beam of light. Charles Messier cataloged it on August 3, 1764, as entry 31 in his list of objects a serious observer should not confuse with comets.

M31 is easily found with the small telescope. Located just above Nu Andromeda, it is easily visible as a greenish glow with a low-power, wide-field eyepiece. It takes a long-time-exposure photograph to reveal the stars that make up M31, but you will be able to trace the extent of the galaxy from its luminous central area to the points where it fades out to the east and west of center.

On a night with excellent transparency you should be able to detect M31's companion galaxy just south of the galaxy's nucleus. This smaller galaxy, M32, shines at magnitude 9. Also visible with the small telescope is another companion, M110.

Located to the northeast of the nucleus, this galaxy is slightly dimmer at magnitude 10.8, but larger than M32.

Located in the constellation Canes Venatici—at RA 13h29m, Dec +47°12′—is M51—really two objects joined together. The primary object is M51 at magnitude 8.4. It is a face-on spiral discovered by Messier in October 1773. At the tip of one of the spiral arms lies NGC 5195, a companion galaxy discovered by Mechain in 1781. Although Messier never saw the companion, it shows faintly in most small telescopes available today—because even today's smallest instruments are superior to those available to some of astronomy's greats. A small telescope will show the galaxy as a hazy spot with a bright nucleus. The spiral structure becomes more apparent with larger instruments.

M81 is another spiral galaxy, similar to M31, found in the constellation Ursa Major at RA 9h56m, Dec +69°04′. This galaxy is magnitude 6.9 and similar to M31 in type. It is turned so that it almost faces us. Discovered by Frederick Bode in December 1774, it was described as a "nebulous patch, more or less round with a dense nucleus in the middle." When Messier added this object to his catalog in February 1781, he described it as a bright little oval. Many of the observers of M81 are impressed with the brightness of the object. Rev. T. W. Webb, writing during the mid-nineteenth century, said M81 is "bright with a vivid nucleus."

This is another object where the observer is treated to a two-for-one sight. In the same field is another galaxy, M82. Classed as an irregular galaxy, M82 seems to be undergoing the throes of an explosion beyond imagination.

Other galaxies suitable for observation with a small telescope are listed in Appendix F.

GALAXIES, SUPERNOVAE, AND THE AMATEUR

The study of galaxies is pretty much in the hands of professional astronomers and their big telescopes and sophisticated equipment. There is one area in which the amateur can make a contribution, however: the systematic monitoring of galaxies for supernovae. The requirements for this project do include a telescope over 8 inches in aperture, but with the explosive growth of large-aperture amateur instruments, many amateurs already have the tools they need.

A supernova is the massive explosion of a star accompanied by its rapid, dramatic increase in brightness. Stars in another galaxy are not usually visible in the amateur telescope. But a supernova can be seen as a point of light that brightens very rapidly, usually in one or two days, and then decreases in brightness over a relatively short period of time. The trick in maintaining such a program is knowing what the galaxy you are watching looks like without any extra points of light.

This is not as hard as it may seem. First off, the probability that a galaxy you are monitoring for the first time will have a supernova is a bit rare. These events don't occur every day. A supernova is a very infrequent event—only two have been recorded in our own Milky Way galaxy in almost 1,000 years.

But people do find these elusive explosions elsewhere outside our galaxy. For example, Rev. Robert Evans of Coonabarabran, New South Wales, Australia, is the current leader in visual supernova discoveries. Since early 1999, he has discovered 32 supernovae visually, mostly with backyard telescopes. Three of these were found visually with a professional telescope, and a few others, plus a comet, have been found photographically in a pro-am search he was briefly involved in. "Thirty-two visual discoveries is a record by quite a margin," said Evans. "Many amateurs are now getting involved in searching for supernovae using CCDs, and are having very good successes." In fact, he adds that because the competition is now much steeper than 10 years ago, it probably takes more than 10,000 negative galaxy observations (on the average) to score an supernova discovery—which shows the dedication of the supernova hunters.

If you can find a copy of *Hubble's Atlas of Galaxies*, by all means grab it. Not only is it a beautiful book, it also shows the appearance of many galaxies in their normal, nonsupernova state. All you have to do is become familiar with a galaxy, something you will do by repeated observation anyway—and notice if something seems to change.

It is best to limit your monitoring to galaxies that show you a large portion of their face, usually spiral galaxies. In these types of galaxies, you have the best chance of success. The Messier catalog gives you about 26 potential candidates for a systematic supernova patrol. They are all spiral galaxies and have an integrated magnitude brighter than 10. The entire procedure is simple: Check each galaxy once every clear night. If you see something you think may be a supernova, contact the Central Bureau of Astronomical Telegrams (similar to reporting a new comet)—but only after checking and rechecking to make sure of your discovery.

DRAWING THE DEEP SKY

We've talked about this before, but drawing deep-sky objects is different from drawing the Moon and planets. No two deep-sky objects are alike, even if they are both of the same type. You can record descriptions of the objects in your log, but take a good look at M42 and really try to describe it. It's hard to do. And it's not that written descriptions aren't important—they are. They allow you to improve your observational skills and look at objects with a critical, analytic eye. But if written descriptions can do that, a drawing of a deep-sky object will help develop the eye even faster.

Unlike planetary drawing, there are no preprinted forms you can use to sketch a deep-sky object. You need a plain piece of white paper with a circle on it. The circle is to keep you honest, so to speak, and indicates the boundaries of your field of view. The size of the circle is up to you. Some observers use a 2-inch circle; some, a 3-inch circle. Use whatever size makes you comfortable. Then you need some sharp pencils—the kind used in drafting with different hardnesses—an eraser, and a stick of graphite to start sketching.

Get comfortable at your scope. Then look at the object you want to draw with two or three different eyepieces. Use the eyepiece that shows you the most detail for your sketching. Begin by filling in the background stars around the object. Be careful about the placement of the brighter stars in the field; work from the center out to the edges of the field. Next add any little asterisms that catch your eye. Including them can be helpful in making a quick identification of an object you don't know.

If you are drawing an open cluster, carefully place each star in its appropriate place. Open clusters are tough because there are usually many stars to record, so they can take a while to draw. If you are drawing an object with a nebulosity, rub some of the graphite onto the paper and then use your thumb to smear it around into the approximate shape of the object. The first smear will give you a gray tone. If you want to darken it, wet your thumb and rub it again until you have the desired tone. To lighten it, use your eraser to skim off a bit. If you are recording the appearance of a globular cluster, rub on some graphite before you add the stars that are visible in your telescope. Then add the stars you can resolve. This gives the globular the same 3-D effect it has in the eyepiece.

A planetary nebula such as the Ring nebula is easy to draw. It has a definite boundary you can easily outline. If you detect shadings in the nebula, add them by using the flat or the point of the graphite stick. Add the central star only if you can see it.

Your drawing should be an accurate representation of what you see in the eyepiece. Don't add anything because you know about where it should be on the drawing. Galaxies, like planetaries, show a definite outline, so draw that first and then work in toward the nucleus. If you are cursed with a large telescope, you also will have to add details of your view like dust lanes and bright patches.

Bright nebulae will give you the most trouble. With their wide range of brightness and subtle shadings across their surface, it is often best to smear down a uniform layer of graphite and make it darker or lighter as you move out from the center of the object. When you are finished with your sketch, you will have an important addition to your hopefully growing observation log.

14

OBSERVING
THE SUN

The influence of this eminent body . . . is so great . . .
that it becomes almost a duty to study the operations
which are called on upon the solar surface.

SIR WILLIAM HERSCHEL

Sometimes it's hard to understand astrophysicists. They spend their nights trying to fathom the life cycle of a small, average star. In the process they stay up long hours and rob themselves of sleep—when all they have to do is roll out of bed, open their shade, and look at the greatest stellar laboratory imaginable. (Fig. 14-1)

The Sun is a special object for the amateur astronomer. It must be treated with respect and dealt with carefully. *It is the only astronomical object that can physically harm you.* Galileo learned the hard way. Turning his telescope on the dim image of the Sun at sunset, he was shocked to learn that the intensity of the Sun didn't really diminish. After his first look, he was blind for a week. Even after his sight returned, the after-effects followed him for the remainder of his life in the form of deteriorating eyesight. Because he took one short peek at the unshielded Sun, Galileo died blind.

WARNING! Never look directly at the Sun through a telescope or binoculars of any kind without proper protection.

Do you remember years ago, when they sold novelty lighters that would light a cigarette using a mirror to focus the rays of the Sun? Well, a binocular or telescope objective does the job much better. The intensity of the light focused by even a small objective lens or mirror produces sufficient heat to set a solid block of wood on fire. You do not want to put your eye *anywhere near* that kind of heat.

245

FIGURE 14-1 *View of the Earth directly between the Sun and the Apollo 12 spacecraft during its journey home from the Moon. (NASA)*

TELESCOPES, FILTERS, AND THE SUN

Most telescopes purchased at the local department store come with a small, screw-on type of filter that supposedly reduces the intensity of the Sun's image. If your telescope came with one of these filters, the best way to use it is to grasp it firmly in your hand and *throw it as far away from you as possible*. These so-called filters *are not safe*. Remember, the optical system of a small telescope, even the finder, will produce an intense amount of heat. If you put this little filter at the focus of that heat, it is going to crack. Even a tiny crack will allow enough heat into your eye to cause permanent damage.

There are only two safe ways to observe the Sun. First, you can reduce the intensity of the solar image *before* it passes through the telescope's optical train. Reflective aperture filters are made of a thin layer of metal, usually Inconel, that is evaporated over an element of optical glass. This in turn is aluminized on both the inside and outside layer so that most of the Sun's light is reflected *away* from the telescope; it allows only about 0.1 percent of the Sun's normal brightness to enter the optical train. By reflecting the brightness of the Sun away from the telescope, the heat caused by the intense solar image is prevented from entering your telescope's optical system. Using an aperture filter is the *only* way for completely safe, direct viewing of the Sun.

A full-aperture filter is the same size as your objective. It completely covers the front end of your telescope and looks like a mirrored end cap. A subdiameter aperture filter also fits over the end of your instrument, but it is mounted with an aperture mask similar to the one used for planetary observing and the filter surface is smaller. Full-aperture filters are fine for small retractors and instruments with an unobstructed light path. Subdiameter filters work best with reflectors and cata-dioptrics with a large secondary obstruction.

Aperture filters are made by a number of companies (see Appendix K for a complete list). They can be made with two types of filter material, each with different transmission ranges. Glass filters made with Inconel produce a yellow image of the Sun. Filters can also be made with a thin sheet of Mylar, which gives the Sun a blue image. Because the eye is less sensitive to the blue end of the spectrum, the Mylar filters tend to lose small details on the solar surface.

A subdiameter filter is preferred over full-diameter filters in some instances. The same considerations given to atmospheric conditions affecting seeing at night apply to observing the Sun during the day. The same air cells that cause erratic seeing while observing the planets and the Moon can affect your observations of the Sun. This is further complicated during the day because the Sun is heating the air, which rises as "thermals" off everything: roofs, pavement, garden sheds, and so on. When you use the subdiameter aperture filter, you will look through fewer of these cells—and the image improves.

If you use a reflective aperture filter of any kind, there is one caution that bears passing along. *Cover your finder scope or get an additional filter for the finder*. The finder, after all, is a telescope in its own right and will act to intensify the heat from the Sun. Although you may not be foolish enough to look through it, someone else may not have your common sense. How are you supposed to find the Sun if your finder is covered? It's really quite simple. Just look at the shadow of your telescope (Fig. 14-2).

FIGURE 14-2 *Shadow of a telescope to align to the Sun.*

Move the telescope around until you get the smallest shadow, then you know the scope is pointed at the Sun.

WARNING! Never look directly at the Sun through a telescope,
finder scope, or binoculars of any kind without proper protection.

The other safe method of solar observation is the *projection method*. This is both the easiest and the safest method for observing the Sun (Fig. 14-3). In the projection method, the Sun's image passes through your telescope and is projected onto a flat white surface, such as a sheet of paper or a card. You observe this image by facing the projected image with your back to Sun.

One thing to remember when using the projection method is that the Sun is unfiltered, and therefore undiminished in intensity, as it passes through your telescope. This limits the type of eyepiece you can use to project the solar image. Any eyepiece that contains multiple elements cemented together, like a plossl, orthoscopic, or RKE, cannot be used. They are going to be placed where the solar image comes to a focus, right at the point of the image's greatest intensity. Because of this, the cement that binds the elements together will be subjected to intense heat and will melt, rendering the eyepiece useless. It is better to use an eyepiece with single, noncemented elements like a ramsden or huygens. (This is the reason for the suggestion in Chapter 5—to tuck the eyepieces away in a drawer.) SCTs should be used with an aperture mask even when using the projection method. The heat builds up inside the sealed tube and can damage the cement holding the secondary to its mount.

The actual construction of a projection system is easy. First determine the distance between the projection screen and the eyepiece. Do this by holding a white

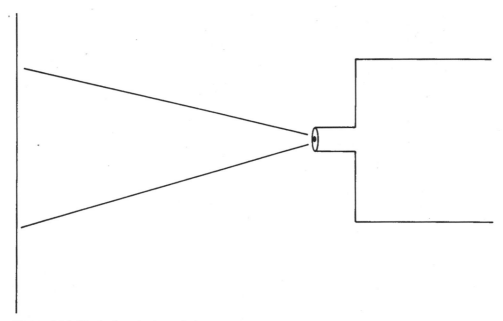

FIGURE 14-3 *The basic projection technique.*

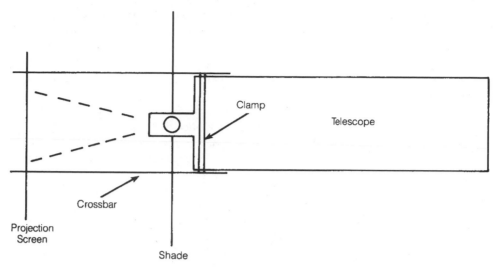

FIGURE 14-4 *A projection screen should be shaded from the glare of the Sun.*

sheet of paper about 12 inches away from the eyepiece and carefully focusing the image of the Sun on the paper. The image should fill the paper and allow you to easily see any visible details. If it's not sharp enough, move the paper back and forth until you get a sharp, detailed image, then measure that distance. Now use a thin strip of wood or metal (metal bends easier) to make the support for the projection screen. Attach it to the telescope with any convenient type of clamp. Large hose clamps work nicely for this.

Because the image of the Sun grows dimmer as it moves away from the eyepiece and onto the screen, it can be overwhelmed by the glare of the Sun itself. If the optical train of the telescope and eyepiece form a straight line, use a second piece of paper or posterboard to shade the screen (Fig. 14-4). Even with a newtonian it is not a bad idea to shade the screen with an oversized piece of posterboard. Bend the posterboard so that it acts as a shade.

Both the reflective aperture filter and the projection method allow you to observe the Sun in normal white light. Observations can also be made in *monochromatic light* using special devices that filter the Sun's light, admitting only one wavelength. This wavelength is usually the *hydrogen-alpha band*, and it can afford some absolutely spectacular views of the Sun that are missed with white light. Equipment to see this band of light, in the form of hydrogen-alpha filter systems, is expensive. So we will confine our look at the Sun to those features you can see in white light.

THE SOLAR LANDSCAPE

With your reflective aperture filter or projection screen in place, still remember our warning.

> *WARNING! Never look directly at the Sun through a telescope or binoculars of any kind without proper protection.*

The Sun looks peaceful enough, but looks can be deceiving. Solar activity takes place in 11-year periods called *solar cycles*. During periods of *solar minimum*, the lowest phase of activity, the Sun can seem a pretty dull place. But for every minimum there is a corresponding *solar maximum* during which the sun really heats up. Throughout the solar cycle, there are a number of features that the amateur can observe and record easily.

At first glance the bright solar surface, or *photosphere*, seems flat and featureless. But if you watch and the seeing is excellent, you will notice that the seemingly featureless photosphere is covered with tiny splotches, almost like the graininess of a photograph when viewed up close. This graininess is caused by small cells called *granules*, which range in size from 1 to 5 arcseconds across, with an average size of 2.5 arcseconds.

Granule visibility is the measure from which solar observers can determine seeing quality. When the seeing is excellent, better than 1 arcsecond, the granules are easily resolved and seen. If the granules are seen as mottled or spotty, the seeing is rated at 1 to 2 arcseconds or good. If the granules appear to pop in and out of view, the seeing is 2 to 5 arcseconds and is rated as fair. Anything beyond this 5-arcsecond limit is poor seeing.

The most obvious features of the solar disk are the dark *sunspots* (Fig. 14-5 and 14-6), relatively cool features that float on the photosphere. If you look carefully, you will see that a sunspot is more than just a spot. The dark, central area is called the *umbra*. It is usually, but not always, surrounded by a lighter, but still dark, *penumbra*. The penumbra itself can show a lot of detail. If the seeing is good, you can see dark

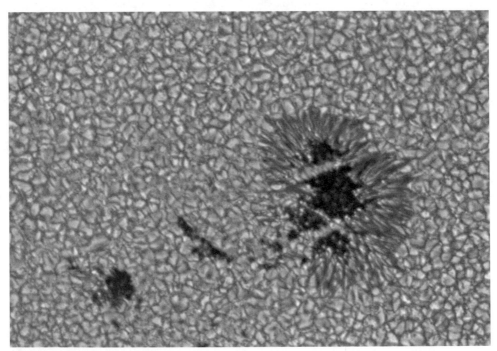

FIGURE 14-5 *Dark spots on the Sun are sunspots, which usually last for several days. The darker section is the umbra; the lighter section is the penumbra. (NASA)*

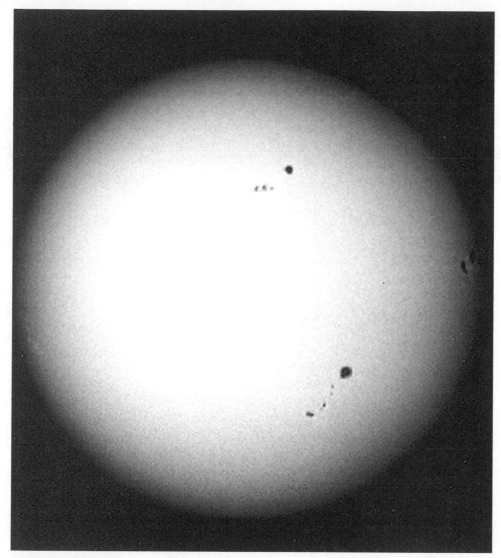

FIGURE 14-6 *Sunspots on the Sun. (R. M. Sandy and NASA)*

areas called *filaments* and brighter areas called *grains*. These may be overlaid with a system of *fibrils*, thin dark lines radiating outward from the umbra. Very small sunspots with no penumbra are called *pores*.

Sunspots tend to form in pairs and cluster in magnetically related groups as they move across the photosphere. Sunspots and sunspot groups come in a wide variety of shapes and sizes. To simplify and standardize descriptions of these groups the *Zurich sunspot classification system* was devised. Sunspot activity follows the approximate 11-year solar cycle. At solar minimum there may be few, if any, sunspots, while at solar maximum the surface can appear riddled with them. The periods of solar maximum include around 1990, 2000, and so on.

Sunspots rotate with the Sun, completing a period in a little over 27 days (although rotation varies a bit from equator to poles). Seen near sunspots are bright patches called *faculae*. Because they are so bright, faculae have little contrast against the almost equally bright background of the photosphere. When we look at the center of the solar disk, we are looking straight through the Sun's atmosphere, called the *corona*. When we look at the limb of the Sun, we are looking through more of the corona, with the limb darker than the rest of the solar disk. This *limb darkening* gives you the best chance to see the faculae; there is much more contrast against the darker limb than against the brighter regions away from the limb.

Another type of bright patch often crosses the sunspot's umbra. These are called *light bridges*. The appearance of a light bridge in a sunspot usually indicates that the spot is about to break apart and dissolve into the photosphere.

Finally, there is the ultimate bright spot seen near a well-developed group of sunspots. *White light flares*, a very rare phenomenon, with only about 12 recorded since the invention of the telescope, can appear as a sudden brightening in or near the sunspot group. They are so bright that they stand out clearly from the bright photosphere, but they only last a brief time.

CONTRIBUTIONS FROM THE AMATEUR

Accurate counts of sunspots are a reliable and important gauge of where the Sun is during the solar cycle. The AAVSO Solar Division collects and correlates sunspot counts from amateurs around the United States and the world. Counting sunspots is an easy and scientifically necessary activity that any amateur can do. To count sunspots, set up your telescope with its filtering or projection equipment and use an eyepiece that allows you to see the entire solar image. Then count the groups you see.

Groups are usually quite obvious. If you see one, lonely little sunspot off by itself, it also must be counted as a group. Don't trust your memory—write the final total down in your log immediately after counting the groups. Next, switch to a higher magnification and scan the photosphere for any small spots not associated with the groups. Again, record this number in your log. Finally, return to each group and carefully count the number of spots contained in each.

When your observation is complete you will have three sets of numbers: the number of groups, the number of single, isolated spots, and the number of spots in the groups. To determine your own daily count, or the day's *American relative spot numbers* (RA), multiply the number of groups by 10 and then add the number of spots. Remember to count each isolated sunspot as *both* a group and a spot. For example, if you count 5 groups of spots, 2 isolated spots, and 13 spots within the 5 groups, your RA would be 85. This is determined by adding 5 groups plus 2 isolated spots to get 7 groups ($7 \times 10 = 70$); then add 13 group spots plus 2 isolated spots for 15 spots: $70 + 15 = 85$.

It is best to try running a count for awhile to determine your accuracy. Each month sunspot numbers for previous months are compiled by such groups as the AAVSO (through their monthly AAVSO Solar Bulletin) and the Sunspot Index Data Center in Brussels, Belgium. Check your relative sunspot numbers against these in order to determine your accuracy. If you are close to the published figures, you should contact AAVSO to request information about joining the Solar Division. You can also pass your early data along to AAVSO for them to analyze.

While the AAVSO Solar Division concentrates on obtaining accurate American relative sunspot numbers, the ALPO Solar Section (ALPOSS) is concerned with the morphology, or the form and structure, of solar phenomena. To better study and analyze these events, ALPOSS compiles a daily visual record of the solar disk. If you're interested, this means drawing something, and solar drawings are among the easiest astronomical drawings to make (Fig. 14-7).

ALPOSS wants daily, full-disk drawings of the Sun. To make the drawings, you first have to determine the inclination of the solar axis relative to Earth (Fig. 14-8). An accurate representation of the solar poles is necessary for an accurate drawing. In your log book, draw a circle to represent the Sun. Mark off the poles with "ticks" and designate north and south for orientation. Using an eyepiece that allows you to see the entire solar disk, mark the positions of each spot or group on the disk. Use dots and dashes to indicate the outlines of the features. Be careful to accurately draw the placement, shape, and size with relation to the other groups on the photosphere and the solar disk itself.

Now go back and fill in the details in the individual groups. Darken the areas to indicate a spot's umbra, paying attention to its proportion with relation to the other

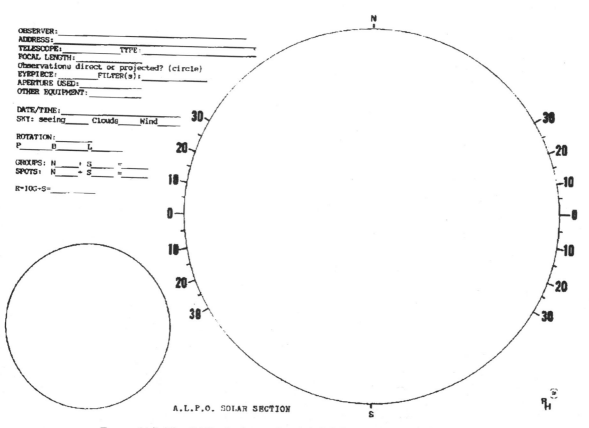

FIGURE 14-7 *The ALPO solar form—the whole disk drawing form. (Courtesy ALPO)*

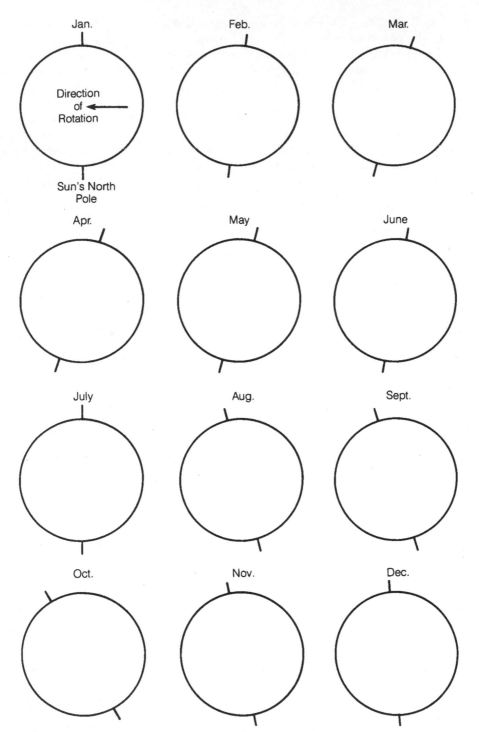

Direction of Rotation

Sun's North Pole

FIGURE 14-8 *The solar wobble. The sun wobbles like a top as the Earth circles it. This shows the location of the Sun's north and south pole throughout the year.*

spots in the group. Outline the penumbra's boundary, which need not be shaded. Solar projection observers have a bit of an edge with whole disk drawings. A sheet of paper can be placed on the projection screen and the detail traced directly off the projected image. The final drawing can be transferred to the back of the ALPOSS report form.

You can also include detailed drawings of individual spots and groups. First, draw an accurate outline of the umbra region and darken it. Then outline the penumbra, shading it to indicate the varying intensities by using light shading for light areas and darker shading for darker areas (but not as dark as the umbra). Be sure to represent the more apparent filaments and bright knots you see in the penumbra. Also be careful in your depiction of the size of the group in relation to the background you are using. Most observers tend to make the regions too large. Again, projection users have a distinct advantage because they are working from an earlier traced projected image.

SOLAR ECLIPSES

One of the most awe-inspiring sights in nature is a total solar eclipse—when the Moon comes between the Earth and the Sun (the Moon is always in the new phase during a solar eclipse) (Fig. 14-9). The slow darkening of the Earth's landscape,

FIGURE 14-9 *Total solar eclipse on February 26, 1998, taken by satellite. (The Exploratorium and NASA)*

caused by the advance of the Moon across the face of the Sun, is eerie. The abrupt cutoff of light at totality brings out features that are invisible during normal observing. And if the sun "overlaps" the moon's figure as it reaches totality, a feature called Bailey's beads shows up—actually the sunlight shining through the mountains and valleys of the Moon. (Fig. 14-10)

During the total eclipse, the ghostly, shimmering corona is often visible, as are the red *prominences* and sometimes *flares* (Figs. 14-11, 14-12, and 14-13)—huge eruptions extending hundreds of thousands of miles into space. Totality, however, doesn't last very long—at most, about 7 minutes—so it's difficult to view these huge eruptions.

Unfortunately, total solar eclipses are rare events and are confined to very narrow strips on the Earth's surface. Predictions of solar eclipses for the year can be found in the *Astronomical Almanac*, the RCAS's *Observer's Handbook*, and even places such as astronomy magazines and Guy Ottewell's annual *Astronomical Calendar*.

More common solar eclipses occur when the Moon is a bit farther away from the Earth—and thus, the Moon's shadow does not quite reach our planet's surface. During these *annular eclipses*, the Moon is seen to be outlined by the bright disk of the Sun.

During a total eclipse, an amateur can do a few things that can't normally be done. First, you can search for comets almost right up to the now dark solar disk. Because totality is so short, comet searching is best done with a camera. A series of

FIGURE 14-10 *A total solar eclipse with Bailey's beads demonstrates the sunlight passing through the valleys and mountains on the Moon. (NASA)*

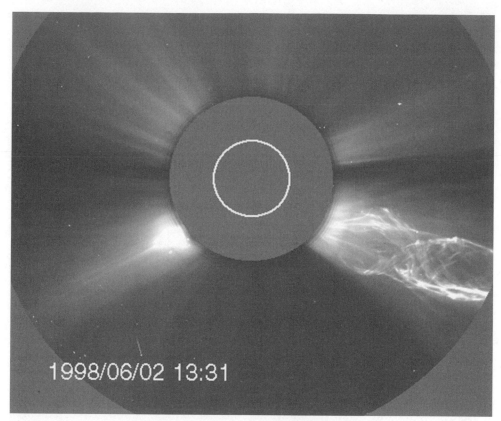

1998/06/02 13:31

FIGURE 14-11 *An enormous eruptive prominence from the SOHO satellite. (NASA)*

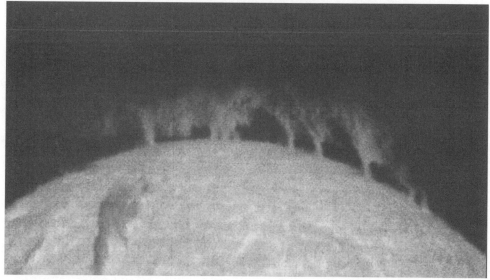

FIGURE 14-12 *Dense clouds of material—prominences—suspended above the surface of the Sun by loops of magnetic fields. (NASA)*

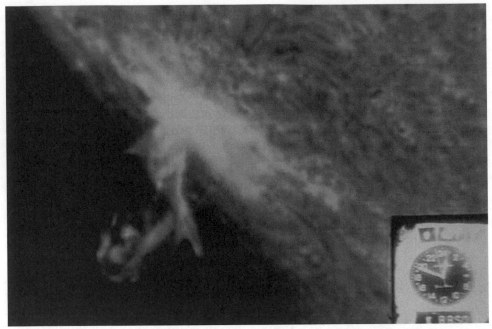

FIGURE 14-13 *A flare blasting off the Sun's limb in just a few minutes time. (Big Bear Solar Observatory and NASA)*

exposures may net a comet as it rounds the Sun at perihelion. You can also easily find elusive Mercury and Venus. If they are not too close to the Sun, you can keep them in telescopic view as totality ends, then observe them against the brightening background of the sky.

The other type of Moon-Sun-Earth interaction is a lunar eclipse—the result of the Earth coming between the Sun and the Moon (the Moon is always at its full phase during a lunar eclipse). A *total lunar eclipse* occurs when the Moon's orbit takes it directly into the Earth's umbra shadow. The result is a blackening out of the Moon. But it is never truly black—the refraction from the Earth's atmosphere shines on the Moon (called earthshine), making the Moon look reddish in color. A *partial lunar eclipse*, in which the Earth's penumbra or umbra is only partially shadowing the Moon's surface, is more common.

You won't always be fortunate enough to catch a lunar eclipse. Many times, the continent on which you reside is facing away from the Moon during the eclipse. In other words, it's often daylight where you are when the eclipse occurs. And just as you do for the solar eclipse, check monthly magazines, such as *Astronomy* or *Sky & Telescope*, or the *Astronomical Almanac* for the times of lunar eclipse events.

15

THE AMATEUR EXPERIENCE

*Sharing a dazzling view of Jupiter, the double cluster,
or M31 with a new observer or a child is beyond reward.*

JEFFREY L. HUNT, Cosmic Connections Inc.

So now you have a telescope and some eyepieces. You've had the scope out, and you've had a look at a representative sample of what the universe has to offer. Amateur astronomers go through stages in their development. You became interested in astronomy and you've done something about it. You've read this book (a wise choice) and purchased or made a telescope. You are a sky watcher. As a sky watcher you observe for the sheer pleasure of probing the wonders of the universe. There is no particular pattern to your observing, but that's all right. Remember, this is a hobby, something do in your leisure time. You may have a full-time job, school, or something that makes more important demands on your time. This is fine—and there is nothing wrong with being a "part-time" sky watcher.

But after a while, the simple thrill of just sky watching can wear a bit thin. The first clue is when the telescope comes out less often and stays out for less time. When you get around to dragging out the scope, the views still give you that sense of wonder, but now something is missing. Simply getting the telescope out of the house has turned into a chore. Observing itself seems to have lost its charm and become more like a job. The process of going out each night to observe becomes less attractive because the "I've seen it all" syndrome rears its ugly head. As a result, the telescope spends more time in the closet or basement and less time outside under the stars.

As more telescopes are retired and more amateur astronomers become "former" amateur astronomers, the problem is probably simple boredom. Unfortunately, many amateurs react to this growing boredom by putting an ad in the paper or one of the astronomy magazines and selling the telescope. They solve the problem by eliminating it, as you can see if you read the classified section in *Astronomy* magazine or get into a telescope exchange site on the Internet: listing after listing of "For Sale: Schmidt-Cassegrain 8-inch, only used once" (or twice, or whatever), followed by a ridiculously low price. Each time one of these ads appears, the community of amateur astronomers has probably lost another of its members to boredom.

It is much better to give your observing some sort of focus—or something to concentrate on so that you never have a chance to get bored. Somewhere along the line an observer realizes that in addition to the sheer joy of sky watching, he or she may be able to make a small contribution to science. When the light dawns, amateurs take the first step toward becoming serious observers. Now their observations take a more orderly tack—and they are out at their telescopes on a regular basis, though maybe not every night. They make their observations regularly, systematically, and with some concrete goal in mind.

For some, it's observing variable stars; for others it is a systematic program directed at a particular planet such as Jupiter or the long task of searching for a comet. The joy of the sky watcher is still there, but it is focused by a desire to learn something more about what is being observed. Some amateurs make the switch from sky watcher to serious observer early in their career; others take years to make the transition.

No law that says you must become a serious observer. Some of us never make the change, enjoying our random views of the universe for our entire observing lives. But it is interesting to note that serious observers enjoy what they are doing as much as the beginning sky watcher—if not more. At a public star party you can't tell them from the casual sky watcher. Most serious observers have combined their scientific curiosity with the simple joys of sky watching.

THE STARS AND THE SOLITARY OBSERVER

For the most part, amateur astronomy is a solitary avocation. The long hours an amateur spends at the telescope are usually spent alone. Part of the lure of amateur astronomy is, we think, the solitude that it offers the observer. It gives many people an outlet, a chance to escape the pressures of everyday life. But no amateur can consider himself or herself a serious observer and function in a vacuum. That is not how science works—not even amateur science. To get the most out of amateur astronomy you have to establish lines of communication with other amateurs.

Many amateurs gravitate to groups that encompass their observing specialties. Lunar and planetary observers have the Association of Lunar and Planetary Observers (ALPO)—and with the advent of the Internet, more and more people are joining the ranks. Founded in 1947 by Walter Haas, ALPO is divided into sections devoted to a particular member of the solar system. Some of the sections are further divided into specialized program areas. For example, the ALPO Lunar Section has programs devoted to lunar transient phenomena, selected lunar areas, eclipses and photometry, and a lunar dome survey. Each section or subsection is headed by a recorder, who has the responsibility of collecting and analyzing the observations sent in by the section's members. For a complete list of the ALPO sections, see Appendix C.

ALPO publishes a journal, *The Journal of the Association of Lunar and Planetary Observers* (JALPO), affectionately known as "The Strolling Astronomer," four times a year. It contains a wealth of information directed at the planetary and lunar observer. ALPO also has a section for the new observer, "The Lunar and Planetary Training Program," that can help a beginner develop his or her observational skills.

Variable star observers have the American Association of Variable Star Observers (AAVSO). Founded in 1911, AAVSO is the clearinghouse for amateur observations of variable stars. AAVSO also is divided into subsections, each with its own area of specialization. Eclipsing binaries, novae and supernovae, Cepheid variables, and the Sun are some of the subsections. AAVSO also puts out a number of publications to help the amateur develop the skills needed to become a working member of the amateur scientific community.

YOUR LOCAL ASTRONOMY CLUB

Probably the best source of assistance for the beginning observer is the local astronomy club. Thousands of amateur astronomers across the United States are members of local astronomy clubs. From memberships of over 1,000 to small groups of a dozen or so, these clubs play an important part in the growth of the amateur.

Membership in your local club can give you a forum for comparing and discussing observations and techniques. Your membership can also bring additional benefits such as access to equipment that you could never possess on your own.

One of the most exciting events held by local groups is the public star party. If you haven't bought a telescope yet, these functions can give you an introduction to telescopes you can't get anywhere else. You can compare, side by side, the images produced by reflectors, retractors, SCTs, and a host of weird-looking beasts that reach out to the stars and bring them to Earth. You can talk to active amateurs who can help with information on making or buying your telescope. They can also explain the goals of the local group—and even how you can join the club.

Don't feel that you have to be an astronomical whiz to join your local group. Many members join with little or no knowledge about the field. Amateur astronomy is a growth experience. Members of your local club can help you along the path that best suits your needs and desires.

SHARING THE EXPERIENCE

To be truly appreciated, astronomy must be shared. Whether it's through lectures to a senior citizens group, in a line of people waiting to see Saturn at the club's star party, or with the kids down the block, astronomy holds no greater thrill for the amateur than when it is shared.

Mike's experience with his astronomical society was typical:

"Each month the Chicago Astronomical Society co-hosted a sky watch with Triton College in one of Chicago's suburbs. When I started to attend these I did so out of curiosity and, to be honest, I didn't expect much. The first time I went I was amazed to see over 200 people calmly stand in line at the members' telescopes for a peek at Saturn or the Moon. Because I didn't bring my telescope I had a chance to wander around and listen to some of the comments and discussions going on between the visitors and members.

"What I heard was an education in itself. Questions were asked and answered on almost every aspect of telescopes and astronomy that you could imagine. The common "what power is this thing?" was asked and answered by a patient member who explained about the importance of image quality over magnification. I heard a couple of kids, they couldn't have been more than 10, ask about the latest theories about lunar crater formation."

With one telescope trained on Saturn, its owner explained that in addition to the rings of the planet the person at the eyepiece could see some of the planet's moons. He explained that the brightest moon, Titan, was visible at about five o'clock in the field. A housewife with a little boy in tow took a look and was puzzled at what she saw. "How many satellites are visible?" she asked. The telescope operator promptly replied that with his instrument a person should be able to see five of the moons. She looked back and asked, "Is that one at about 3 o'clock, close to the rings?" The operator quickly checked and confirmed that Rhea was indeed visible and admitted that he had overlooked it. You don't have to be an amateur astronomer to have an astronomer's eye.

"My turn came to man the telescopes when a member wanted a short break" said Mike. "I quickly found myself at a 10-inch reflector. As each person came to the eyepiece I told them that they would be looking at Saturn and to look for as long as they wanted. The line progressed smoothly with each person uttering a comment on the planet's beauty. A little girl moved along with her mother and when it was her turn she slowly mounted the step ladder to the eyepiece. She looked for a moment, said she couldn't see anything, and started down. I told her to take her time and she turned back to the eyepiece. Watching her fidget at the eyepiece I knew that the image of the planet had not hit her eye. Then all of a sudden she froze in place. The look on her face told me she had found Saturn. She looked at her mother and then turned back to the eyepiece where she remained for a long time. All she could say was Wow!"

Jeff Hunt is right. And you'll out find, too—a reaction like that is the best reward astronomy can offer.

WHY WE TRY

Probably more than anything, we as amateurs try to discover the universe in our own backyard—as in the proverbial question about why people climb Mt. Everest—because it's there. But in reality, becoming an amateur astronomer means more.

According to Don Parker, executive director of the Amateur Lunar and Planetary Observers (ALPO), each year, amateur astronomy keeps getting better. Telescopes and eyepieces, not to mention charts and data collection, are becoming more standardized. And with the Internet, more amateurs are staying in touch across town and around the world.

The contributions to science make us all swell with pride, too, whether it is someone we know or another amateur across the country. Take for example Mars: The multitude of probes recently sent to the red planet hasn't slowed down the contributions by amateur astronomer. "In fact," says Parker, "even with the Mars Global Surveyor, we still watch Mars. The craft is in a polar orbit—so if the rest of the planet exploded, we'd never know. That's where the amateur astronomers come in—to watch the rest of the planet." And ALPO also has the only long-term weather data for Mars, including trends and clouds, and even the annual shrinking and growing of the polar caps.

Whether you're watching Saturn, trying to split double stars, or just keeping track of your own favorite stars, amateur astronomers do make a difference. You don't have to contribute to science either. You can just revel in the joy of your own personal astronomical discoveries—even pretend what it was like at the dawn of the telescope. Astronomy is meant to be fun and adventurous.

Take your eyes, binoculars, or telescopes out and view the heavens in any way you see fit. Remember, it's a beautiful universe out there. Go out and discover it.

INTERNET LINKS OF INTEREST

Like the new CCD technology, the Internet is opening up a whole new world for amateur astronomers. Not only is it easier to find information, but quick and easy communication between amateurs—even with professionals—is now possible. (One note: It may be easier to contact people via the Internet, but don't overuse the privilege. Don't bother people with a steady stream of e-mail. After all, it is just like regular mail, but faster. Be polite in your requests and responses.) Some amateur astronomers have also set up their own home pages. If you decide to do the same, try linking your site to others with an astronomical theme.

Here are dozens of links to astronomy sites on the Internet. There are plenty more—and the numbers will no doubt increase in the future. The Internet is definitely a place where one thing leads to another.

ASTRONOMY GROUPS

Abrams Planetarium
http://www.pa.msu.edu/abrams/SkyWatchersPage/Index.html

American Association of Variable Star Observers (AAVSO)
http://www.aavso.org

American Astronomical Society
http://www.aas.org/ (mostly professionals)

Association of Lunar and Planetary Observers (ALPO)
http://www.lpl.arizona.edu/alpo/

Astronomical League
http://www.astroleague.org/

Astronomical Society of the Pacific
http://aspsky.org/

British Astronomical Association
http://www.ast.cam.ac.uk/~baa/

European Space Agency
http://www.esa.int/

Hubble Space Telescope Images by Subject
http://oposite.stsci.edu/pubinfo/subject.html#SolarSystem

International Amateur-Professional Photoelectric Photometry (I.A.P.P.P.)
http://www.iappp.vanderbilt.edu/

International Dark-Sky Association
http://www.darksky.org/ida/ida_2/index.html

International Occultation Timing Association (IOTA)
http://www.anomalies.com/iotaweb/

International Small Telescope Cooperative (professional and amateur)
http://www.astro.fit.edu/istec/

Lick Observatory
http://syborg.ucolick.org/lickobs/index.html

New Star Gazers Page
http://www.calweb.com/~dmurry/index.htm

The Planetary Society
http://planetary.org/

Royal Astronomical Society
http://www.worldserver.pipex.com/ras/

Small Radio Telescope Project from Haystack Observatory
http://fourier.haystack.mit.edu/menuSRT.html

Society for Popular Astronomy
http://www.u-net.com/ph/spa/

Society of Amateur Radio Astronomers
http://www.bambi.net/sara.html

Space Telescope Science Institute
http://www.stsci.edu/

Stellafane Home Page
http://www.stellafane.com/

Sunspot Index Data Center
http://www.oma.be/KSB-ORB/SIDC/

United States Naval Observatory
http://aa.usno.navy.mil/AA/

The Webb Society
http://www.webbsociety.org./wbg.html

The Woman Astronomer
http://www.erols.com/njastro/twa/

MISCELLANEOUS LINKS

Astronomy Magazine (http://www2.astronomy.com/astro/)
A fine site sponsored by Astronomy magazine; it includes their Deep-Sky
Online, a place for deep-sky watchers.

Astronomy Online *(http://www.rt66.com/~breeden/astro/index.html)*
A great place to find information on astronomy—including an observer's check-
list to plan for a night of astronomy with or without a telescope.

Astroweb/Astronomy Web Resources *(http://www.stsci.edu/net-resources.html)*
Extensive list of astronomy-related Web links.

Bushnell Observatory *(http://www.bushnell.com/observatory/observhome.html)*
This site is brought to you by Bushnell optics and offers a sky calendar and
sundry tips on astronomical observing.

"Comets" Homepage *(http://encke.jpl.nasa.gov)*
This comets page is maintained at JPL by Charles Morris, one of the world's
most active comet observers.

Hale-Bopp Comet Page *(http://www.halebopp.com)*
Even though this comet is now out of naked-eye visibility, there are still some
great pieces of information out there on comets—not to mention that larger
telescopes can catch a glimpse of the now-dim comet. (Amateurs could still
catch the comet until 1999.)

The Moon *(http://seds.lpl.arizona.edu/billa/tnp/luna.html)*
The Moon according to the lunar and planetary group in Arizona, with many
more links to information on our only satellite.

Virtual Reality Moon Phase Pictures *(http://tycho.usno.navy.mil/vphase.html)*
Find the phase of the Moon for any date from 1800 to 2199 a.d.

NASA's Amateur Astronomy *(http://www.hq.nasa.gov/office/oss/amateur.htm)*
NASA's site to help amateur astronomers.

Professional and Amateur Astronomy Pages *(http://www.hal-pc.org/~malel/astron.html)*
Links to all sorts of astronomy pages for amateurs and professionals.

Radio-Sky Publishing *(http://radiosky.com/)*
P.O. Box 3552
Louisville, KY 40201-3552
This group has many helpful hints for those interested in radio astronomy.

Sky Online *(http://www.skypub.com/)*
Online astronomical news from the publisher of *Sky and Telescope* and *CCD Astronomy* magazines.

Astronomy Software and CD-ROMs

Not all amateur astronomers use a computer. But for those who do, here are a few astronomical software programs and CD-ROMs to check out. (These are in no way endorsements of these products—just suggestions of what is available to the amateur astronomer with a computer; and there are many more—just put in the keywords "astronomical software" into a search engine on the Internet). Most of these can be purchased through an on-line astronomy store or via astronomical catalogues, such as those distributed by Sky Publishing, Inc. or the Astronomical Society of the Pacific *(http://aspsky.org/)*:

Buil-Thouvenot Atlas CD-ROM
Sky Publishing Corp.
49 Bay State Road
Cambridge, MA 02138
http://www.skypub.com/

A decade's worth of sky images taken with CCD technology are in this version of an atlas; there are more than 4,600 pictures by astro imagers Christian Buil and Eric Thouvenot.

Celestia 2000 CD-ROM (via catalog)

A new CD-ROM from the European Space Agency, containing data from the Hipparcos and Tycho catalogs in compressed binary format.

DigitalSky Voice
Telescope Control System Software
Astro-Physics, Inc.
11250 Forest Hills Road

Rockford, IL 61115
815-282-1513

Uses voice-activation technology to run the telescope.

MegaStar
Willmann-Bell, Inc.
P.O. Box 35025
Richmond, Virginia 23235
http://www.willbell.com

A star atlas covering stars, deep-sky objects, asteroids, etc., for Windows 3.1, 95, and NT.

The Hubble Library of Electronic Picturebooks
STScI
Johns Hopkins University
Baltimore, MD

A collection of space images, artwork, and explanations (including a section on the planets by Barnes-Svarney) put out by the Space Telescope Science Institute.

Pises Atlas-Prism98
Astronomical Society
Montpellier, France

Deep-sky objects from this astronomical society's observatory, with 2,000 objects.

RealSky (via catalog)

This 18 CD-ROM set (8 for the northern celestial pole and 10 for the southern) is for the serious observer, and shows you how the sky really appears through a telescope, with 1- to 2-arc resolution; stars down to 19th or 20th magnitude are also shown.

3-D Tour of the Solar System (by P. Schenk, D. Gwynn, and J. Tutor)
Lunar and Planetary Institute
Houston, Texas

With a CD-ROM and 3-D glasses, this CD takes the viewer through a tour of our solar system.

TheSky
Astronomy Software
Software Bisque
912 Twelfth Street
Golden, CO 80401
http://www.bisque.com

This software is fast and easy to use, simulating the sky from any location (in CD-ROM and a Macintosh version).

Sky Catalogue 2000.0 (via catalog)

Software rendition of Hirshfeld, Sinnott, and Ochsenbein's two-volume catalog.

Skymap Pro 4.0 (via catalog)

This software uses a sophisticated planetarium and map-drawing capabilities with information on millions of objects.

Sky Tools
CapellaSoft
P.O. Box 1182
Cloudcroft, NM 88317
http://www.skyhound.com

A CD-ROM package that offers observational planning, charting, and logging.

Skywatching (David Levy and Robert Burnham) (via catalog)

An easy-to-use introduction to all things astronomical, and a CD-ROM of the book.

Starchart III (via catalog)

An advanced planetarium program.

Starry Night Deluxe
http://www.siennasoft.com

A realistic astronomy program.

U.S. Naval Observatory's Multi Year Computer Almanac, (MICA) 1990-1005
Willmann-Bell, Inc.
P.O. Box 35025
Richmond, VA 23235
http://www.willbell.com

A tool to use when "pretty" sky maps are not enough; it allows you to calculate data for user-specified locations and times.

SELECTED GROUPS FOR THE AMATEUR ASTRONOMER

VARIABLE STARS

AAVSO

The American Association of Variable Star Observers (AAVSO) collects and analyzes information on the brightness changes of variable stars. Information about joining AAVSO may be obtained from

> AAVSO
> 25 Birch St.
> Cambridge, MA 02138
> Director: Dr. Janet A. Mattei

Certain operations of AAVSO are coordinated outside AAVSO Headquarters under the authority of the director, including the following committees and divisions (of which the chairs operate voluntarily): Charge-Coupled Device (CCD); Photoelectric Photometry (PEP); Eclipsing Binary Stars; Solar Division; Nova Search; Supernova Search; RR Lyrae Stars; New Charts Committee; and the Telescope Committee. Each division has certain responsibilities within the group. For example, the AAVSO Solar Division collects information on sunspot activity in the form of sunspot counts. The AAVSO Eclipsing Binary Committee is concerned with gaining accurate estimates of light increase and decrease from eclipsing binaries on the right of their predicted eclipse. (Estimates are made every 10 minutes or so for a period of three or four hours on the night of the eclipse.) And the AAVSO novae and supernovae divisions seek out these explosive events. (The nova searches are carried out using nothing more than an atlas and a pair of 7×50 binoculars, but because the supernova searches are carried out by examining faint galaxies, a telescope of at least medium aperture is required.)

THE SOLAR SYSTEM

ALPO

The goal of the Association of Lunar and Planetary Observers (ALPO) is to gather information on each member of our solar system. Each section is concerned with a specific member of the solar system. Most of the sections publish newsletters and information handbooks separate from the Association's journal (*Strolling Astronomer*). There are also sections on instruments, Mercury/Venus transits, and computing. Direct any inquiries about joining ALPO to the membership secretary.

Executive Director

Donald C. Parker
12911 Lerida Street
Coral Gables, FL 33156

Membership Secretary

Harry Jamieson
ALPO Membership Secretary
P.O. Box 171302
Memphis, TN 38187-1302

Moon

Coordinator
William M. Dembowski
ALPO Coordinator, Lunar
Topographical Studies
219 Old Bedford Pike
Windber, PA 15963

Coordinator, Transient Phenomena
David O. Darling
416 W. Wilson St.
Sun Prairie, WI 53590-2114

Coordinator, Selected Areas
Julius L. Benton, Jr.
Associates in Astronomy
305 Surrey Rd.
Savannah, GA 31410

Coordinator, Lunar Dome Survey
Harry D. Jamieson, Membership
Secretary and Treasurer
P.O. Box 171302
Memphis, TN 38187-1302

Sun

Solar Section, Acting Coordinator
Richard Hill
Lunar and Planetary Lab
University of Arizona
Tucson, AZ 85721

Acting Assistant Coordinator
Jeff Medkeff
Rockland Observatory
3602 Trevino Drive
Sierra Vista, AZ 85635-5262

Mercury

Acting Coordinator
Harry Pulley
532 Whitelaw Road
Guelph, Ontario, Canada N1K
1A2

Venus

Coordinator
Julius L. Benton, Jr.
Associates in Astronomy
305 Surrey Road
Savannah, GA 31410

Mars

STAFF

Daniel M. Troiani (Coordinator; all
observations; U.S. correspondence;
Martian Chronicle mailings, IMP

Alert Notices, ALPO Mars
Observing Kit)
629 Verona Ct.
Schaumburg, IL 60193

Dr. Donald Parker (Assistant
Coordinator)
12911 Lerida Street
Coral Gables, FL 33156

Daniel P. Joyce (Assistant Coordinator, Martian Chronicles Editor)
6203 N. Keeler Avenue
Chicago, IL 60646

Jeff D. Beish (Acting Assistant
Coordinator)
14522 Bisbee Ct.
Woodbridge, VA 22193

Jim Bell (Mars Section Advisor)
Cornell University
Department of Astronomy
Center for Radiophysics and Space
Research
424 Space Sciences Building
Ithaca, NY 14853-6801

Minor Planets

Coordinator
Frederick Pilcher
Illinois College
Jacksonville, IL 62650

Lawrence S. Garrett
206 River Rd.
Fairfax, VT 05454

Jupiter

Acting Coordinator
David J. Lehman
6734 N. Farris
Fresno, CA 93711

John McAnally
(Assistant Coordinator, Transit
Timings)
2124 Wooded Acres
Waco, TX 76710

Craig MacDougal
(Assistant Coordinator)
2602 E 98th Ave
Tampa, FL 33612

Damian Peach
(Assistant Coordinator)
237 Hillington Square
Greyfriars House
Kings Lynn
Norfolk PE30 5HX

John E. Westfall
(Assistant Coordinator, Galilean
Satellites; address under "Editor")

Professor Augustin Sanchez-Lavega
(Assistant Coordinator; Scientific
Advisor)
Departmento Fisica Aplicada I
E.T.S. Ingenieros
Universidad del Pais Vasco
Bilbao, Spain

Sanjay Limaye (Assistant
Coordinator; Scientific Advisor)
University of Wisconsin
Space Science and Engineering
Center
Atmospheric Oceanic and Space
Science Bld. 1017
1225 W. Dayton St.
Madison, WI 53706

Saturn

Coordinator
Julius L. Benton, Jr.
Associates in Astronomy
305 Surrey Road
Savannah, GA 31410

Remote Planets

Coordinator
Richard W. Schmude
Gordon College
Division of Natural Sciences and
Nursing
419 College Dr.
Barnesville, GA 30204

Comets

Coordinator
Don E. Machholz
P.O. Box 1716
Colfax, CA 95713

Assistant Coordinator
James V. Scotti
Lunar and Planetery Lab
Univ. of Arizona
Tucson, AZ 85721

Meteors

Coordinator
Robert D. Lunsford
161 Vance St.
Chula Vista, CA 91910

Assistant Coordinator
Mark A. Davis
1054 Anna Knapp Blvd., Apt. 32H
Mt. Pleasant, SC 29429

Supernovae

The International Supernovae Network (ISN) helps to maintain contact and share information among supernovae enthusiasts—both amateur and professional—worldwide. They can be reached mainly via their Web site, at *http://www.super-novae.net/isn.htm.*

Amateur Radio Astronomy

The Society of Amateur Radio Astronomers (SARA) is a dedicated group of people who formed an international society to learn, trade technical information and do their own observations of the radio sky.

President
Mike Gingell
113 Trotters Ridge Drive
Raleigh, NC 27614
(919) 847-4779
gingell@vnet.net

To join:
Kenneth Watts
2427 Burgundy Drive
Birmingham, AL 35244
(205) 987-0815
kjw@wwisp.com

CCDs

INTERNATIONAL AMATEUR-PROFESSIONAL PHOTOELECTRIC PHOTOMETRY (I.A.P.P.P.)

International Amateur-Professional Photoelectric Photometry (I.A.P.P.P.) for CCD and photometry was formed in June 1980 to increase the astronomical research collaboration among amateurs, students, and professional astronomers—and this includes possible publication for all participants. Membership includes a subscription to the quarterly journal, *I.A.P.P.P. Communications.* Contact

A. M. Heiser
Dyer Observatory
1000 Oman Drive
Brentwood, TN 37027-4143

THE WEBB SOCIETY

The Webb Society *(http://www.webbsociety.org./)* is interested in the study of double stars and deep-sky objects, including clusters, nebulae, and galaxies. Although the Society is based in England, interested observers in the United States may contact

> R. W. Argyle in England
> E-mail: *rwa@ast.cam.ac.uk*

Those in the United States and Canada interested in applying for membership, contact

> John Isles
> Secretary/Treasurer
> 11105 Tremont Lane
> Plymouth, MI 48170
> Phone and fax: (734) 451-7069
> E-mail: *jisles@voyager.net*

Dark Skies

THE INTERNATIONAL DARK-SKY ASSOCIATION

The International Dark-Sky Association (IDA) is a group of astronomers, both amateur and professional, and private individuals who have banded together to do something about the growing problem of light pollution. The aim of the group is public education, promotion of the use of lighting fixtures that preserve the dark sides above the nation's cities, monitoring of sky brightness, and promotion of existing local ordinances that protect the night sky.

Information concerning membership and the resources IDA has available for amateur astronomers and membership can be obtained by contacting

> International Dark-Sky Association, Inc.
> 3225 North First Avenue
> Tucson, AZ 85719-2103
> (502) 293-3198
> *http://www.darksky.org*

PLANETARY OPPOSITIONS AND SOLAR AND LUNAR ECLIPSES

UPCOMING PLANETARY OPPOSITIONS

YEAR	DATE	DEC (DEGREES)	APPARENT DIAMETER (ARCSECONDS)	MAG
Mars				
1999	APR 24	-11	16.18	-1.4
Jupiter				
1999	OCT 23	+10	49.70	-2.5
2000	NOV 28	+20	48.63	-2.4
Saturn				
1999	NOV 6	+13	20.31	+0.0
2000	NOV 19	+17	20.49	-0.1

Indicates Earth passing through Saturn's ring plane.

ECLIPSES

ECLIPSES OF THE SUN

1999
February 16: annular solar eclipse
August 11: total solar eclipse

2000
February 5: partial solar eclipse
July 1: partial solar eclipse
July 31: partial solar eclipse
December 25: partial solar eclipse

2001
June 21: total solar eclipse
December 14: annular solar eclipse

ECLIPSES OF THE MOON

1999
January 31: penumbral lunar eclipse
July 28: partial lunar eclipse

2000
January 21: total lunar eclipse
July 16: total lunar eclipse

2001
January 9: total lunar eclipse
July 5: partial lunar eclipse
December 30: penumbral lunar eclipse

THE MESSIER CATALOG

COORDINATES FOR EPOCH 2000

M	NGC	RA HR	DEC MIN	DEG	MIN	CONST	MAG	TYPE
1	1952	05	34.5	+22	01	TAU	8.4	DI
2	7089	21	33.5	-00	49	AQR	6.5	GB
3	5272	13	42.2	+28	23	CVn	6.4	GB
4	6121	16	23.6	-26	32	SCO	5.9	GB
5	5904	15	18.6	+02	05	SER	5.8	GB
6	6405	17	40.1	-32	13	SCO	4.2	OC
7	6475	17	53.9	-34	49	SCO	3.3	OC
8	6523	18	03.8	-24	23	SGR	5.8	DI
9	6333	17	19.2	-18	31	OPH	7.9	GB
10	6254	16	57.1	-04	06	OPH	6.6	GB
11	6705	18	51.1	-06	16	SCT	5.8	OC
12	6218	16	47.2	-01	57	OPH	6.6	GB
13	6205	16	41.7	+36	28	HER	5.9	GB
14	6402	17	37.6	-03	15	OPH	7.6	GB
15	7078	21	30.0	+12	10	PEG	6.4	GB
16	6611	18	18.8	-13	47	SER	6.0	OC
17	6618	18	20.8	-16	11	SGR	7.0	DI
18	6613	18	19.9	-17	08	SGR	6.9	OC
19	6273	17	02.6	-26	16	OPH	7.2	GB
20	6514	18	02.6	-23	02	SGR	8.5	DI
21	6531	18	04.6	-22	30	SGR	5.9	OC
22	6656	18	36.4	-23	54	SGR	5.1	GB
23	6494	17	56.8	-19	01	SGR	5.5	OC
24								

M	NGC	RA HR	DEC MIN	DEG	MIN	CONST	MAG	TYPE
25		18	31.6	-19	15	SGR	4.6	OC
26	6694	18	45.2	-09	24	SCT	8.0	OC
27	6853	19	59.6	+22	43	VUL	8.1	PL
28	6626	18	24.5	-24	52	SGR	6.9	GB
29	6913	20	23.9	+38	32	CYG	6.6	OC
30	7099	21	40.4	-23	11	CAP	7.5	GB
31	224	00	42.7	+41	16	AND	3.4	G-S
32	221	00	42.7	+40	52	AND	8.2	G-E
33	598	01	33.9	+30	39	TRI	5.7	G-S
34	1039	02	42.0	+42	47	PER	5.2	OC
35	2168	06	08.9	+24	20	GEM	5.1	OC
36	1960	05	36.1	+34	08	AUR	6.0	OC
37	2099	05	52.4	+32	33	AUR	5.6	OC
38	1912	05	28.7	+35	50	AUR	6.4	OC
39	7092	21	32.2	+48	26	CYG	4.6	OC
40								
41	2287	06	47.0	-20	44	CMa	4.5	OC
42	1976	05	35.4	-05	27	ORI	4.0	DI
43	1982	05	35.6	-05	16	ORI	9.0	DI
44	2631	08	40.1	+19	59	CNC	3.1	OC
45		03	47.0	+24	07	TAU	1.2	OC
46	2437	07	41.8	-14	49	PUP	6.1	OC
47	2422	07	36.6	-14	30	PUP	4.4	OC
48	2548	08	13.8	-05	48	HYA	5.8	OC
49	4472	12	29.8	+08	00	VIR	8.4	G-E
50	2323	07	03.2	-08	20	MON	5.9	OC
51	5194	13	29.9	+47	12	CVn	8.1	G-S
52	7654	23	24.2	+61	35	CAS	6.9	OC
53	5024	13	12.9	+18	10	COM	7.7	GB
54	6715	18	55.1	-30	29	SGR	7.7	GB
55	6809	19	40.0	-30	58	SGR	7.0	GB
56	6779	19	16.6	+30	11	LYR	8.2	GB
57	6720	18	53.6	+33	02	LYR	9.0	PL
58	4579	12	37.7	+11	49	VIR	9.8	G-S
59	4621	12	42.0	+11	39	VIR	9.8	G-E
60	4649	12	43.7	+11	33	VIR	8.8	G-E
61	4303	12	21.9	+04	28	VIR	9.7	G-S
62	6266	17	01.2	-30	07	OPH	6.6	GB
63	5055	13	15.8	+42	02	CVn	8.6	G-S
64	4826	12	56.7	+21	41	COM	8.5	G-S
65	3623	11	18.9	+13	05	LEO	9.3	G-S
66	3627	11	20.2	+12	59	LEO	9.0	G-S
67	2682	08	50.4	+11	49	CNC	6.9	OC

M	NGC	RA HR	DEC MIN	DEG	MIN	CONST	MAG	TYPE
68	4590	12	39.5	-26	45	HYA	8.2	GB
69	6637	18	31.4	-32	21	SGR	7.7	GB
70	6681	18	43.2	-32	18	SGR	8.1	GB
71	6838	19	53.8	+18	47	SGE	8.3	GB
72	6981	20	53.5	-12	32	AQR	9.4	GB
73								
74	628	01	36.7	+15	47	PSC	9.2	G-S
75	6864	20	06.1	-21	55	SGR	8.6	GB
76	6501	01	42.4	+51	34	PER	11.5	PL
77	1068	02	42.7	-00	01	CET	8.8	G-S
78	2068	05	46.7	+00	03	ORI	8.0	DI
79	1904	05	24.5	-24	33	LEP	8.0	GB
80	6093	16	17.0	-22	59	SCO	7.2	GB
81	3031	09	55.6	+69	04	UMa	6.8	G-S
82	3034	09	55.8	+69	41	UMa	8.4	G-Ir
83	5236	13	37.0	-29	52	HYA	7.6	G-S
84	4374	12	25.1	+12	53	VIR	9.3	G-E
85	4382	12	25.4	+18	11	COM	9.2	G-E
86	4406	12	26.2	+12	57	VIR	9.2	G-E
87	4486	12	30.8	+12	24	VIR	8.6	G-E
88	4501	12	32.0	+14	25	COM	9.5	G-S
89	4552	12	35.7	+12	33	VIR	9.8	G-E
90	4569	12	36.8	+13	10	VIR	9.5	G-E
91	4548	12	35.4	+14	30	COM	10.2	G-S
92	6341	17	17.1	+43	08	HER	6.5	GB
93	2447	07	44.6	-23	52	PUP	6.2	OC
94	4736	12	50.9	+41	07	CVn	8.1	G-S
95	3351	10	44.0	+11	42	LEO	9.7	G-S
96	3368	10	46.8	+11	49	LEO	9.2	G-S
97	3587	11	14.8	+55	01	UMa	11.2	PL
98	4192	12	13.8	+14	54	COM	10.1	G-S
99	4254	12	18.8	+14	25	COM	9.8	G-S
100	4321	12	22.9	+15	49	COM	9.4	G-S
101	5457	14	03.2	+54	21	UMa	7.7	G-S
102								
103	581	01	33.2	+60	42	CAS	7.4	OC
104	4594	12	40.0	-11	37	VIR	8.3	G-S
105	3379	10	47.8	+12	35	LEO	9.3	G-E
106	4258	12	19.0	+47	18	CVn	8.3	G-S
107	6171	16	32.5	-13	03	OPH	8.1	GB
108	3556	11	11.5	+55	40	UMa	10.0	G-S
109	3992	11	57.6	+53	23	UMa	9.8	G-S
110	205	00	40.4	+41	41	AND	8.0	G-E

EXPLANATION OF TYPES

DI, diffuse nebula   DI
GB, globular cluster   GB
OC, open cluster   OC
PL, planetary nebula   PL
G-S, galaxy-spiral   G-S
G-E, galaxy-elliptical   G-E
G-Ir, galaxy-irregular   G-Ir

COMMON NAMES FOR SOME OF THE MESSIER OBJECTS

M1	Crab nebula
M8	Lagoon nebula
M11	Wild Duck cluster
M13	Great Hercules cluster
M17	Omega nebula
M20	Triffid nebula
M27	Dumbbell nebula
M31	Andromeda galaxy
M42	Orion nebula
M44	Beehive cluster or Praesepe
M45	Pleiades
M51	Whirlpool galaxy
M57	Ring nebula
M64	Blackeye galaxy
M76	Little Dumbbell nebula
M97	Owl nebula
M104	Sombrero galaxy

THE MISSING MESSIERS

I'm sure that you noticed there are a few objects listed with only a number, nothing else. These four objects were mistakes in identification made by Messier when he compiled his original catalog.

M24 is a bright area of the Milky Way in Sagittarius.
M40 is a wide double star in Ursa Major.
M73 is a small asterism in Aquarius.
M102 is a misidentification of M101.

DEEP-SKY OBJECTS
FOR THE TELESCOPE

COORDINATES FOR EPOCH 2000

NGC	RA HR	MIN	DEC DEG	MIN	CONST	MAG	CLASS
Open Clusters							
457	01	19.1	+58	20	CAS	6.4	I-3-R
663	01	46.0	+61	15	CAS	7.1	III-2-M
752	01	57.8	+37	41	AND	5.7	III-1-M
1857	05	20.2	+39	21	AUR	7.0	II-2-M
2192	06	15.2	+39	51	AUR	10.9	III-1-P
2324	07	04.2	+01	03	MON	8.4	II-2-R
2362	07	18.8	-24	57	CMa	4.1	I-3-P
2432	07	40.9	-19	05	PUP	10.2	II-1-P
2509	08	00.7	-19	04	PUP	9.3	II-1-P
6530	18	04.8	-24	20	SGR	4.6	II-2-M
6611	18	18.8	-13	47	SER	6.0	II-3-M
6819	19	41.3	+40	11	CYG	7.3	II-1-R
7142	21	45.9	+65	48	CEP	9.3	II-2-R
Globular Clusters							
5897	15	17.4	-21	01	LIB	8.5	11
Planetary Nebulae							
2392	07	29.2	+20	55	GEM	9.9	irregular disk
3242	10	24.8	-18	38	HYA	8.6	ring structure

NGC	RA HR	MIN	DEC DEG	MIN	CONST	MAG	CLASS
Planetary Nebulae							
6210	16	44.5	+23	49	HER	9.3	smooth disk
6543	17	58.6	+66	38	DRA	8.8	irregular disk
7009	21	04.2	-11	22	AQR	8.3	ring structure
7293	22	29.6	-20	48	AQR		ring structure
Bright Nebulae							
1432	03	46.1	+23	47	TAU		very faint-R
1788	05	06.9	-03	21	ORI		bright-R
1975	05	35.4	-04	41	ORI		bright-E + R
2261	06	39.2	+08	44	MON		bright-E + R
2359	07	18.6	-13	12	CMa		very faint-E
7000	20	58.8	+44	20	CYG		faint-E
Galaxies							
253	00	47.6	-25	17	SCL	7.0	spiral
278	00	52.1	+47	33	CAS	10.8	elliptical
2403	07	36.9	+65	36	CAM	8.9	spiral
4449	12	28.2	+44	06	CVn	9.8	irregular
4494	12	31.4	+25	47	COM	9.8	elliptical
4565	12	36.3	+25	59	COM	10.3	spiral

DOUBLE STARS FOR THE TELESCOPE

What follows is a list of 118 double and multiple stars. Many of them are easily accessible by a small telescope; others may be beyond an observer's instrument or observing conditions. Each pair is easy to locate. They are all listed by name in the latest edition of *Norton's Star Atlas and Reference Handbook*.

Each star is listed by name, with STR indicating stars from F. G. W. Struve's catalog and OSTR denoting stars from his son Otto's work. Following the name is the constellation in which the star is found, the separation of the pair, and the magnitudes of the components. The position angle (PA) is listed last, with triple stars designated by a T after the PA.

NAME		SEPARATION (ARCSEC)	MAGNITUDES A	B	POSITION ANGLE
OSTR 514	AND	5.3	7.0	9.5	168
STR 24	AND	5.2	7.0	8.0	249
STR 79	AND	7.8	6.0	7.0	193
59	AND	16.6	6.5	7.0	35
STR 245	AND	11.2	7.0	8.0	294
STR 2985	AND	15.3	7.0	8.0	254
STR 3042	AND	5.3	7.0	7.0	88
29	AQR	3.9	7.0	7.0	244
41	AQR	5.0	6.0	7.5	114
STR 3008	AQR	4.0	7.0	8.0	175
107	AQR	6.6	5.5	6.5	137
5	AQL	13.0	5.5	7.5	121
STR 2426	AQL	17.1	7.0	8.0	259
STR 2446	AQL	9.6	6.5	8.5	153 (T)
			34.5	9.5	341
STR 2489	AQL	8.2	6.5	9.5	347
STR 2532	AQL	33.7	6.0	10.0	5

Name		Separation (arcsec)	Magnitudes A	B	Position Angle
STR 2613	AQL	4.0	7.0	7.5	352
1	ARI	2.7	6.0	7.5	166
Omega	AUR	5.4	5.0	8.0	359
41	AUR	5.5	5.0	7.0	356
Kappa	BOO	13.3	4.5	6.5	236
Pi	BOO	5.6	5.0	6.0	108
Epsilon	BOO	2.9	2.5	5.0	338
39	BOO	2.9	6.0	6.5	45
Xi	BOO	6.9	5.0	7.0	347
STR 419	CAM	3.0	7.0	7.0	75
1	CAM	10.3	5.0	6.0	308
STR 973	CAM	12.5	6.5	7.5	31
STR 1127	CAM	5.3	6.0	8.0	340 (T)
			11.3	9.0	175
Iota	CNC	30.5	4.5	6.5	307
OSTR 195	CNC	9.6	7.5	8.0	139
2	CVn	11.4	5.5	8.0	260
STR 1645	CVn	10.2	7.0	7.5	158
Nu-1	CMa	17.5	6.5	8.0	262
STR 3053	CAS	15.2	6.0	7.5	70
Eta	CAS	11.0	3.5	7.5	297
Iota	CAS	2.2	4.0	7.0	241 (T)
			7.3	8.0	114
Sigma	CAS	3.0	5.5	7.5	326
Kappa	CEP	7.4	4.0	8.0	122
Beta	CEP	13.6	3.5	8.0	250
STR 2819	CEP	12.4	7.5	8.5	57
Xi	CEP	7.6	4.5	6.5	278
Omicron	CEP	3.2	5.5	8.0	214
26	CET	16.0	6.5	9.0	253
66	CET	16.2	6.0	7.5	232
Nu	CET	8.0	5.0	9.5	83
STR 1633	COM	9.0	7.0	7.0	245
24	COM	20.3	5.0	6.5	271
Zeta	CrB	6.3	5.0	6.0	305
OSTR 304	CrB	10.7	6.5	10.0	175
Gamma	CRT	5.2	4.0	9.0	96
Delta	CYG	2.2	3.0	6.5	247
16	CYG	39.0	5.0	5.0	134
17	CYG	26.0	5.0	8.0	70
Psi	CYG	3.1	5.0	7.5	177
52	CYG	6.6	4.0	9.0	67
STR 2725	DEL	5.7	7.5	8.0	9
Gamma	DEL	10.1	4.5	5.0	268

Name		Separation (arcsec)	Magnitudes A	B	Position Angle
Eta	DRA	5.3	3.0	8.0	143
17	DRA	3.2	5.5	6.5	109
Epsilon	EQU	10.6	5.5	7.0	70
20	GEM	20.0	6.0	7.0	210
38	GEM	7.0	5.5	7.5	151
Lambda	GEM	9.6	3.0	10.0	33
Delta	GEM	6.3	3.5	8.0	218
Kappa	GEM	7.0	4.0	10.0	239
Kappa	HER	28.2	5.0	6.0	12
Alpha	HER	4.6	3.0	5.5	110
Delta	HER	8.8	3.0	8.5	241
Rho	HER	4.1	4.5	5.5	317
95	HER	6.3	5.0	5.0	258
100	HER	14.2	6.0	6.0	183
17	HYA	4.3	7.0	7.0	359
Theta	HYA	29.4	4.0	10.0	197
N	HYA	9.1	6.0	6.0	210
54	HYA	8.8	5.0	7.0	126
STR 1360	LEO	14.2	7.5	8.0	242
54	LEO	6.5	4.5	6.5	110
88	LEO	15.4	6.5	8.5	326
18	LIB	19.6	6.0	9.5	39
STR 1962	LIB	11.8	6.5	6.5	189
19	LYN	14.7	5.5	6.5	315
20	LYN	15.1	7.0	7.0	254
Zeta	LYR	43.7	4.5	5.5	150
STR 2470	LYR	13.5	6.5	8.0	271
Eta	LYR	28.3	4.0	8.0	83
8	MON	13.2	4.5	6.5	27
Beta	MON	7.4	5.0	5.5	132
STR 926	MON	11.7	7.5	8.5	289
Zeta	MON	32.0	5.0	10.5	245
19	OPH	23.2	6.0	9.5	89
39	OPH	10.8	6.0	7.0	355
61	OPH	20.6	6.0	6.5	93
STR 627	ORI	21.0	6.5	7.0	260
Rho	ORI	7.0	4.5	8.5	63
23	ORI	32.0	5.0	7.0	28
Lambda	ORI	4.4	4.0	6.0	44
Iota	ORI	11.4	3.0	7.0	141
Theta	PER	18.3	4.0	10.0	303
Eta	PER	28.4	4.0	8.5	301
40	PER	20.0	5.0	9.5	238
Zeta	PER	12.9	3.0	9.5	209

NAME		SEPARATION (ARCSEC)	MAGNITUDES A	B	POSITION ANGLE
Epsilon	PER	8.8	3.0	8.0	9
35	PSC	11.8	6.0	7.5	149
Zeta	PSC	23.6	4.5	5.5	63
Theta	SGE	11.9	6.5	8.5	325
Sigma	SCO	20.0	3.0	9.0	273
Delta	SER	3.9	4.0	5.0	178
6	SER	3.1	5.5	9.5	20
Theta	SER	22.2	4.5	5.0	103
STR 430	TAU	26.1	6.0	9.0	55 (T)
			37.0	10.0	301
Chi	TAU	19.5	5.5	7.5	25
Iota	TRI	3.8	5.0	6.5	72
STR 232	TRI	6.5	7.5	7.5	246
Nu	UMa	7.2	4.0	10.0	147
Gamma	VIR	4.7	3.5	3.5	306
STR 2769	VUL	17.9	6.5	7.5	300

Variable Stars
for the Telescope

Design	Name	Range	Period	Type
*010884	RU CEP	8.2-9.4	109d	SR
015254	U PER	8.1-11.3	321d	LPV
*0214-03	Omicron CETI	3.6-9.1	332d	LPV
021558	S PER	8.6-10.6	?	SR
*0243-12	Z ERI	7.0-8.6	80d	SR
*0247-08	RR ERI	7.4-8.6	97d	SR
032043	Y PER	8.4-10.3	253d	LPV
0455-14	R LEP	6.8-9.8	421d	LPV
*0506-11	RX LEP	5.0-7.0	?	IRR
*052504	CK ORI	5.9-7.1	120d?	SR
053068	S CAM	7.7-11.1	327d	SR
053613	FX ORI	8.2-10.4	720d	SR
*054907	Alpha ORI	0.4-1.3	2070d	SR
072046	Y LYN	7.8-10.3	110d	SR
081112	R CNC	6.8-11.2	361d	LPV
*0824-05	RT HYA	7.5-9.2	253d	SR
085120	T CNC	7.6-10.5	482d	SR
*085211	RT CNC	8.3-9.4	90d	SR
094211	R LEO	5.8-10.0	312d	LPV
*112245	ST UMa	7.7-9.5	81d?	SR
115158	Z Uma	7.9-10.8	196d	SR
*122001	SS VIR	6.8-8.9	355d	LPV
123307	R VIR	6.9-11.5	146d	LPV
123961	S UMa	7.8-11.7	226d	1pv
*131546	V CVn	6.7-8.8	192d	SR
*1324-22	R HYA	4.5-9.5	390d	LPV
142539	V BOO	7.0-12.0	258d	SR
153738	RR CrB	8.4-10.1	60d	SR

DESIGN	NAME	RANGE	PERIOD	TYPE
*155947	X HER	7.5-8.6	95d	SR
1621-12	V OPH	7.5-10.2	298d	LPV
*162542	g HER	5.7-7.2	70d	SR
163172	R UMi	9.1-10.4	324d	SR
164715	S HER	7.6-12.6	307d	LPV
181136	W LYR	7.9-12.2	197d	LPV
*183308	X OPH	6.8-8.8	334d	LPV
*185243	R LYR	3.9-5.0	46d	LPV
190108	R AQL	6.1-11.5	291d	LPV
194048	RT CYG	7.3-11.8	190d	LPV
*200938	RS CYG	7.2-9.0	417d	SR
201647	U CYG	7.2-10.7	462d	LPV
203816	S DEL	8.8-12.0	277d	LPV
*204017	U DEL	7.6-8.9	?	SR
210868	T CEP	6.0-10.3	388d	LPV
*214058	Mu CEP	3.6-5.1	?	SR
231040	TY AND	8.8-10.5	260d	SR
233335	ST AND	8.8-11.1	328d	SR
2338-15	R AQR	6.5-10.3	387d	LPV

*Indicates binocular variable.

*I

CONSTELLATIONS VISIBLE FROM 40° NORTH

Name	Abbreviation	Season Visible
Andromeda	And	Fall
Antila	Ant	Spring
*Aquarius	Aqr	Spring
Aquila	Aql	Summer
*Aries	Ari	Fall
Auriga	Aur	Winter
Bootes	Boo	Summer
Caelum	Cae	Winter
Camelopardalis	Cam	Winter
*Cancer	Cnc	Spring
Cards Venatici	CVn	Spring
Canis Major	CMa	Winter
Canis Minor	CMi	Winter
*Capricornus	Cap	Fall
Cassiopeia	Cas	Fall
Centaurus	Cen	Summer
Cephus	Cep	Fall
Cetus	Cet	Fall
Columba	Col	Winter
Coma Bernices	Com	Spring
Corona Austrais	CrA	Summer
Corona Borealis	CrB	Summer
Corvus	Crv	Spring
Crater	Crt	Spring
Cygnus	Cyg	Summer
Delphinus	Del	Summer
Draco	Dra	Summer
Equuleus	Equ	Fall

Name	Abbreviation	Season Visible
Eridanus	Eri	Winter
Fornax	For	Fall
*Gemini	Gem	Winter
Grus	Gru	Fall
Hercules	Her	Summer
Hydra	Hya	Spring
Lacerta	Lac	Fall
*Leo	Leo	Spring
Leo Minor	LMi	Spring
Lepus	Lep	Winter
*Libra	Lib	Summer
Lupus	Lup	Summer
Lynx	Lyn	Winter
Lyra	Lyr	Summer
Microscopium	Mic	Summer
Monoceros	Mon	Winter
Ophiuchus	Oph	Summer
Orion	Ori	Winter
Pegasus	Peg	Fall
Perseus	Per	Fall
*Pisces	Psc	Fall
Pisces Austrinus	PsA	Fall
Puppis	Pup	Winter
Pyxis	Pyx	Spring
Sagitta	Sge	Summer
*Sagittarius	Sgr	Summer
*Scorpius	Sco	Summer
Sculptor	Scl	Fall
Scutum	Sct	Summer
Serpens	Ser	Summer
Sextans	Sex	Spring
*Taurus	Tau	Winter
Triangulum	Tri	Fall
Ursa Major	UMa	Spring
Ursa Minor	UMi	Summer
*Virgo	Vir	Spring
Vulpecula	Vul	Summer

Indicates zodiacal constellation.

THE GREEK ALPHABET

Alpha	Α
Beta	β
Gamma	γ
Delta	δ
Epsilon	ε
Zeta	ζ
Eta	η
Theta	θ
Iota	ι
Kappa	κ
Lambda	λ
Mu	μ
Nu	ν
Xi	ξ
Omicron	ο
Pi	π
Rho	ρ
Sigma	σ
Tau	τ
Upsilon	υ
Phi	φ
Chi	χ
Psi	ψ
Omega	ω

ASTRONOMICAL SUPPLIERS

Finding good astronomical equipment can be more difficult than finding a magnitude-15 galaxy from the center of New York City. When you look for an equipment manufacturer or dealer, don't look only at price—see what others say about their quality. These groups also can be fountains of information for the new observer. Once you find one, treat them with the respect of a fine eyepiece. Here is a very short list of contacts, some with URL (home page) addresses:

Manufacturers of Telescopes and Accessories
Celestron International
2835 Columbia Street
Torrance, CA 90503
http://www.celestron.com

Manufactures a complete line of refractor, reflector, and catadioptric telescopes and accessories.

Meade Instruments Corporation
6001 Oak Canyon
Irvine, CA 92620
http://www.meade.com

Manufactures a complete line of refractor, reflector, and catadioptric telescopes, optics, and accessories.

Questar Corp.
P.O. Box 59
New Hope, PA 18938
http://www.questar-corp.com/

Manufactures a line of maksutov telescopes.

TeleVue
Dept. A
100 Route 59
Suffern, NY 10901
http://www.televue.com

Manufactures a line of refractors, eyepieces, mountings, and accessories.

DEALERS/TELESCOPE SALES

Adorama
42 West 18th Street
New York, NY 10011
http://www.adoramacamera.com

Offers telescopes and accessories.

Astronomics
2401 Tee Circle
Suites 105/106
Norman, OK 73069

http://www.astronomics.com

Edmund Scientific
Department A991C821
101 E. Gloucester Pike
Barrington, NJ 08007-1380
http://www.edsci.com

Offers all sorts of astronomical and scientific equipment.

Hunt's Photo & Video
100 Main Street
Melrose, MA 02176
http://www.wbhunt.com

Eyepieces, some telescopes, and binoculars.

Orion Telescopes and Binoculars
P.O. Box 1815
Santa Cruz, CA 95061-1815
http://www.telescope.com

Offers several lines of telescopes, accessories, and binoculars.

Pocono Mountain Optics
104 NP 502 Plaza
Moscow, PA 18444

Lines of telescopes and eyepieces—eastern company.

Pocono West Optics
2580 S. Decatur #3
Las Vegas, NV 89102

Lines of telescopes and eyepieces—western company.

Scope City
730 Easy Street
Simi Valley, CA 93065

Buys, sells, trades, and services telescopes and some binoculars.

Starsplitter Telescopes
3228 Rikkard Drive
Thousand Oaks, CA 91362
http://www.starsplitter.com

Offers portable and lightweight telescopes.

ACCESSORIES, FILTERS, AND TELESCOPE-MAKING SUPPLIES

Many amateurs prefer to build their own telescopes from scratch or from commercially produced parts. The following companies supply such things as mirrors, telescope components, and accessories.

Collins Electro Optics LLC
9025 East Kenyon Avenue
Denver, CO 80237
http://www.ceoptics.com

Offers "image intensification" technology for amateur and professional astronomers.

DayStar Filter Company
P.O. Box 1290
Pomona, CA 91769
(714) 591-4673

Specializes in solar filters and nebular filters.

Learning Techniques, Inc.
40 Cameron Avenue
Somerville, MA 02144
(617) 628-1459
http://www.starlab.com/

Lumicon
2111 Research Dr., #5
Livermore, CA 94550
(925) 447-9583

Specializes in telescope accessories, light pollution and nebular filters, and astrophoto accessories.

Tectron Telescopes
2111 Whitfield Park Avenue
Sarasota, FL 34243
http://www.icstars.com/tectron/
Offers telescopes.

Thousand Oaks Optical
Box 248098
Farmington, MI 48332
(805) 491-3642
Specializes in precision full-aperture solar filters.

CCDS

Jim's Mobile Inc.
810 Quail Street, Unit E
Lakewood, CO 80215
Offers a one-man portable CCD machine.

Santa Barbara Instrument Group
P.O. Box 50437
Santa Barbara, CA 93150
http://www.sbig.com
Offers high-quality CCD cameras for amateur astronomers.

BINOCULARS

Bushnell Sports Optics Worldwide
9200 Cody
Overland Park, KS 66214
http://www.bushnell.com/

Fujinon
10 High Point Drive
Wayne, NJ 07470
(201) 633-5600

Swarovski Optik
1 Wholesale Way
Cranston, RI 02920
http://www.swarovskioptik.com/

BOOKS, MAGAZINES, AND BIBLIOGRAPHY

You can never have enough astronomy books and magazines. Here are a few to choose from—and there are more every year:

BOOKS AND MAGAZINES

Abrams Planetarium
Michigan State University
East Lansing, MI 48824

Publishes monthly *Sky Calendar*. (The current issue of the *Sky Calendar* is available over the Internet: *http://www.pa.msu.edu/abrams/SkyCalendar/Index.html*)

Guy Ottewell
Department of Physics and the Astronomical League Universal Workshop
Furman University
Greenville, SC 29613
http://www.kalend.com

This yearly calendar (one of many offerings) is one of the best for easily locating planets, constellations, eclipses, etc.

Cobblestone Publishing
30 Grove Street
Peterborough, NH 03458

Publishes *Odyssey* magazine (monthly), an astronomy magazine for children.

Kalmbach Publishing Co.
21027 Crossroads Circle
P.O. Box 1612
Waukesha, WI 53187
http://www.astronomy.com

Publishes *Astronomy* magazine (monthly).

Sky Publishing Corp.
Box 9111
Belmont, MA 02178-9918

Publishers of astronomical books and *Sky & Telescope* magazine (monthly); *Sky-Watch* (yearly); and *CCD Astronomy,* among many astronomical book offerings.

Willmann-Bell, Inc.
P.O. Box 35025
Richmond, VA 23235
http://www.willbell.com

Publishers of astronomical books and some software.

BIBLIOGRAPHY

GENERAL ASTRONOMY

Burnham, R., Dyer, A., Garfinkle, R., George, M. Kanipe, J., and Levy, D. *Advanced Skywatching*. New York: Time Life (Nature Company Guide), 1997.

Grice, N. *Touch the Stars* (for the visually impaired).

Odenwald, S. *The Astronomy Cafe*. New York: W.H. Freeman, 1998.

AMATEUR GUIDES

Pasachoff, J. M. and Menzel, D. H. *Peterson's Field Guide: A Field Guide to the Stars and Planets*. New York: Houghton Mifflin, 1998.

LEARNING THE SKY

Dickinson, T. *Nightwatch: A Practical Guide to Viewing the Universe*. New York: Firefly Books, 1998.

Dickinson, T. and Dyer, A. *The Backyard Astronomer's Guide*. New York: Camden House Publishing, 1991.

MacRobert. A. M. *Star-Hopping for Backyard Astronomers*. Cambridge, MA: Sky Publishing, 1994.

Schaaf, D. *40 Nights to Knowing the Sky*. New York: Owl Books, 1998.

BINOCULAR ASTRONOMY

Harrington, P. S. *Touring the Universe Through Binoculars*. New York: John Wiley, 1990.

Peltier, L. C. *The Binocular Stargazer*. Waukesha: Kalmbach Publishing, 1995.

SEEING, TELESCOPES, AND OPTICS

Harrington, P. S. *Star Ware: The Amateur Astronomer's Ultimate Guide to Choosing, Buying, and Using Telescopes and Accessories*. New York: John Wiley, 1998.

Ratledge, D. (ed.) *The Art and Science of CCD Astronomy*. New York: Springer Verlag, 1997.

OBSERVING THE MOON: GENERAL

NASA. *Fifty Year Canon of Lunar Eclipses: 1986-2035*. Washington: NASA, 1989.

OBSERVING THE MOON: ATLASES

Rukl, A. *Atlas of the Moon*. Waukesha: Kalmbach Publishing, 1992.

OBSERVING THE PLANETS: GENERAL

Beatty, J. K., Petersen, C. C., and Chalkin, A. *The New Solar System*. Cambridge, MA: Sky Publishing and Cambridge University Press, 1999 (4th edition).

Greeley, R. and Batson, R. *The NASA Atlas of the Solar System*. Cambridge, UK: Cambridge University Press, 1996.

Price, F. W. *The Planet Observer's Handbook*. Cambridge, UK: Cambridge University Press, 1998 (reprint).

COMETS, ASTEROIDS, AND METEORS

Barnes-Svarney, P. *Asteroid: Earth Destroyer or New Frontier?* New York: Plenum Publishing, 1996.

Bone, N. *Meteors*. Cambridge, MA: Sky Publishing, 1994.

Hale, A. *Everybody's Comet: A Layman's Guide to Comet Hale-Bopp*. Silver City, NM: High-Lonesome Books, 1996.

Levy, D. *The Quest for Comets*. New York: Avon Books, 1995.

Yeomans, D. *Comets*. New York: John Wiley, 1991.

OBSERVING DOUBLE STARS

Terrell, D., Mukherjee, J. D., and Wilson, R. E. *Binary Stars: A Pictorial Atlas*. Melbourne, FL: Krieger Publishing Company, 1992.

OBSERVING VARIABLE STARS

Levy, D. *Observing Variable Stars*. Cambridge, UK: Cambridge University Press, 1998 (2nd edition).

Scovil, C. *The AAVSO Star Atlas*. Cambridge, MA: Sky Publishing, 1990 (2nd edition).

OBSERVING DEEP-SKY OBJECTS

Harrington, P. S. *The Deep Sky: An Introduction*. Cambridge, MA: Sky Publishing, 1998.

Kozak, J. T. *Deep-Sky Objects for Binoculars*. Cambridge, MA: Sky Publishing, 1988.

Luginbuhl, C. B. and Skiff, B. A. *Observing Handbook and Catalogue of Deep-Sky Objects*. SIGS Books and Multimedia, 1999.

Petersen, C. C. and Brandt, J. C. *Hubble Vision*. Cambridge, UK: Cambridge University Press, 1998 (2nd edition).

OBSERVING THE SUN

Beck, R., Hilbrecht, H., Reinsch, K., and Volker, P. *Solar Astronomy Handbook*. Richmond, VA: Willmann-Bell, Inc., 1995.

NASA. *Fifty Year Canon of Solar Eclipses: 1986-2035*. Washington, DC: NASA, 1989.

STAR ATLASES AND CATALOGS

Arnold, H., Doherty, P., and Moore, P. *The Photographic Atlas of the Stars*. Waukesha, WI: Kalmbach Publishing, 1997.

Burnham, R. Jr. *Burnham's Celestial Handbook* (volumes 1-3). New York: Dover, 1983.

Dibon-Smith, R. *The Flamsteed Collection*. Clear Skies Publishing, 1998.

Hirshfeld, A., Sinnott, R. W., and Ochsenbein, F. *Sky Catalogue 2000.0, Volume 1: Stars to Magnitude 8.0*. Cambridge, UK: Cambridge University Press, 1992.

Hirshfeld, A. and Sinnott, R. W. *Sky Catalogue 2000.0, Volume 2: Double Stars, Variable Stars, and Nonstellar Objects*. Cambridge, UK: Cambridge University Press, 1992.

Kepple, G. and Sanner, G. *Night Sky Observer's Guide, Volume 1: Autumn and Winter*. Richmond, VA: Willmann-Bell, Inc., 1999.

Kepple, G. and Sanner, G. *Night Sky Observer's Guide, Volume 2: Spring and Summer*. Richmond, VA: Willmann-Bell, Inc., 1999.

Pennington, H. *The Year-Round Messier Marathon Field Guide*. Richmond, VA: Willmann-Bell, Inc., 1998.

Ridpath, I. *Norton's Star Atlas and Reference Handbook*. Reading, MA: Addison-Wesley, 1998 (19th edition).

Sinnott, R. W. and Perryman, M. A. C. *Millennium Star Atlas*. Cambridge, MA: Sky Publishing, 1997. (Based on the data from the European Space Agency's Hipparcos satellite.)

Tirion, W. *The Cambridge Star Atlas*. Cambridge, UK: Cambridge University Press, 1996.

Tirion, W. *Bright Star Atlas 2000.0*. Richmond, VA: Willmann-Bell, Inc., 1990.

Tirion, W. and Sinnott, R. W. *Sky Atlas 2000.0*. Cambridge, MA: Sky Publishing and Cambridge University Press, 1999 (deluxe version, field version, or desk version).

Sinnott, R. (ed.) *NGC 2000.0*.

BOOK CLUBS

Astronomy Book Club
3000 Cindel Drive
Delran, NJ 08370-0001

INDEX

Page numbers followed by *f* indicate figures; page numbers followed by *t* indicate tables.

ABOUT THE AUTHORS

Patricia L. Barnes-Svarney is the author of more than 20 books on popular science for adults and children, as well as numerous articles in such journals as *Popular Science, Air & Space, Astronomy, Final Frontier, Omni, and Ad Astra*. Her extensive background in the physical sciences includes degrees in Geology and Geography, and professional experience in astronomy, geomorphology, and physical oceanography.

Michael R. Porcellino (deceased) was an amateur astronomer for more than 30 years, as well as an active member of the Chicago Astronomical Society, the Association of Lunar and Planetary Observers, and the American Association of Variable Star Observers. In addition to publishing extensively in the field of astronomy, he also scripted training films, manuals, and brochures on the subject.